HOMELAND SECURITY
OPERATIONAL ANALYSIS CENTER

Strategies to Mitigate the Risk to the National Critical Functions Generated by Climate Change

ANDREW LAULAND, LIAM REGAN, SUSAN A. RESETAR, JOIE D. ACOSTA, RAHIM ALI, EDWARD W. CHAN, RICHARD H. DONOHUE, LIISA ECOLA, TIMOTHY R. GULDEN, CHELSEA KOLB, KRISTIN J. LEUSCHNER, MICHELLE E. MIRO, TOBIAS SYTSMA, PATRICIA A. STAPLETON, MICHAEL T. WILSON, CHANDLER SACHS

About This Report

In January 2021, President Joseph Biden issued an executive order directing the Secretary of Homeland Security to "consider the implications of climate change in the Arctic, along our Nation's borders, and to National Critical Functions [NCFs]" (Executive Order 14008, 2021). The 55 NCFs represent vital government and private-sector functions that are essential for the nation's security, economic security, public health, and safety.

The U.S. Department of Homeland Security (DHS) Cybersecurity and Infrastructure Security Agency (CISA) asked the Homeland Security Operational Analysis Center (HSOAC), a federally funded research and development center (FFRDC) operated by the RAND Corporation, to develop a risk management framework and to assess the risk of climate change to higher-vulnerability NCFs. The risk-assessment framework and findings are documented in Miro et al., 2022. This report extends the earlier work to identify and evaluate climate adaptation strategies to address the 25 NCFs identified as at greatest risk from climate change, with an emphasis on strategies that owner-operators of critical infrastructure might implement, including state, local, tribal, and territorial governments and private-sector stakeholders. The findings should be of interest to CISA; multisector groups with a focus on critical infrastructure, such as sector coordinating councils; and sector risk-management agencies and other federal agencies and partners managing risk to U.S. critical infrastructure and to policymakers, research analysts, and other stakeholders interested in addressing the impacts of climate change.

This research was sponsored by CISA and conducted within the Strategy, Policy, and Operations Program of the HSOAC FFRDC.

About the Homeland Security Operational Analysis Center

The Homeland Security Act of 2002 (Section 305 of Public Law 107-296, as codified at 6 U.S.C. § 185), authorizes the Secretary of Homeland Security, acting through the Under Secretary for Science and Technology, to establish one or more FFRDCs to provide independent analysis of homeland security issues. The RAND Corporation operates HSOAC as an FFRDC for DHS under contract HSHQDC-16-D-00007.

The HSOAC FFRDC provides the government with independent and objective analyses and advice in core areas important to the department in support of policy development, decisionmaking, alternative approaches, and new ideas on issues of significance. The HSOAC FFRDC also works with and supports other federal, state, local, tribal, and public- and private-sector organizations that make up the homeland security enterprise. The HSOAC FFRDC's research is undertaken by mutual consent with DHS and is organized as a set of discrete tasks. This report presents the results of research and analysis conducted under 70RCSA21FR0000052, Assessing Risk to the National Critical Functions as a Result of Climate Change.

The results presented in this report do not necessarily reflect official DHS opinion or policy.

For more information on HSOAC, see www.rand.org/hsoac. For more information on this publication, see www.rand.org/t/RRA1645-1.

Acknowledgments

We would like to thank our research partners at the Cybersecurity and Infrastructure Security Agency, especially Paige Morimoto, and our RAND colleague Quentin Hodgson, who contributed to the cyber vulnerability analysis, and the peer reviewers who provided guidance on the development of this report, Krista Romita Grocholski (RAND) and Jordan Fischbach (the Water Institute of the Gulf).

Summary

Issue

In January 2021, President Joseph Biden issued an executive order directing the Secretary of Homeland Security to "consider the implications of climate change in the Arctic, along our Nation's borders, and to National Critical Functions [NCFs]" (Executive Order 14008, 2021). The NCFs represent "the functions of government and the private sector so vital to the United States that their disruption, corruption, or dysfunction would have a debilitating effect on security, national economic security, national public health or safety, or any combination thereof" (Cybersecurity and Infrastructure Security Agency [CISA], undated, p. 1).

To fulfill the objectives of the executive order, CISA asked the Homeland Security Operational Analysis Center, a federally funded research and development center operated by the RAND Corporation, to develop a risk management framework and to assess the risk of climate change to higher-vulnerability NCFs. The results of the team's climate change risk assessment are documented in Miro et al., 2022. The current report extends the earlier work by identifying and evaluating adaptation strategies to mitigate the risks of climate change for the 25 NCFs assessed in the earlier report to be **at moderate risk of disruption on a national scale** from climate change. As noted in the earlier report, "moderate disruption" on a national scale could include major historical weather events, such as Hurricane Katrina (2005), that lead to major disruption on the local or regional scale and could entail significant risks to life and safety, as well as economic loss.

For this report, we sought to understand (1) what adaptation strategies are available to address climate risk to NCFs, (2) how the effectiveness and feasibility of existing adaptation strategies can be assessed, and (3) what tools are available to assist stakeholders with climate adaptation. We focused on strategies that are potentially relevant to owner-operators of critical infrastructure.

Approach

We followed a four-step process using a risk management framework developed in Miro et al., 2022, that provides a comprehensive and structured approach for understanding the sources of climate risk to the NCFs. First, we identified the NCFs that were assessed to be at greatest risk because of climate change in our previous study (Miro et al., 2022), focusing on the "impact pathways" through which climate change might disrupt a particular NCF. Second, we drew on previously published research and subject-matter expert (SME) input to identify frequently cited or well-regarded adaptation strategies.[1] Third, we rated the strategies on several criteria, including effectiveness, feasibility, whether the strategy created or reduced cybersecurity vulnerability, and the strength of evidence supporting the strategy. Fourth, we identified existing tools and aids that could be used to help critical infrastructure owner-operators select strategies to assess in greater detail and, potentially, pursue. We compiled the results of these steps into an Excel spreadsheet which includes the full list of strategies, sources, ratings, and existing tools for each NCF. We also developed a Tableau tool that facilitates exploration and characterization of the database of strategies. We analyzed the data in the spreadsheet to draw conclusions about the availability and quality of strategies for the NCFs included in the study.

---

[1] *Adaptation* is the word most frequently used to describe climate strategies adopted to reduce vulnerability to the impact of climate change—in this case, the vulnerability of specific NCFs. While these are sometimes described as climate risk mitigation strategies, we use *adaptation* throughout this report to differentiate these strategies from climate mitigation strategies, which are adopted to reduce the likelihood of climate change and its effects.

Key Findings

What Adaptation Strategies Are Available to Address Climate Risk to NCFs?

The NCFs provide a useful lens for assessing risk to critical infrastructure from climate change and identifying adaptation strategies to address that risk. To explore this issue, we focused on **impact pathways**, which describe *how* climate change might disrupt the NCF. An NCF might be associated with multiple impact pathways. Each pathway represents a unique combination of an NCF, one of eight climate drivers (drought, extreme cold, extreme heat, flooding, sea-level rise, severe storm systems [nontropical], tropical cyclones and hurricanes, and wildfires), and one of four impact mechanisms (physical damage or disruption, input or resource constraint, workforce shortage, and demand change). A climate adaptation strategy attempts to interrupt the impact pathway for the NCF, either by reducing or eliminating the likelihood that a climate driver will disrupt an NCF or by reducing or eliminating the potential consequences of a climate driver on an NCF.

For example, Figure S.1 shows an impact pathway for the NCF Supply Water, along with two potential adaptation strategies to interrupt the flow of the impact pathway. One strategy would be to *build a flood barrier*, such as a levee, to prevent a flood from reaching a drinking water treatment plant or its supply sources to prevent the physical damage. Another strategy would *secure a backup or alternative water supply source* to reduce a potential disruption in operations due to a flood.

We identified a number of adaptation strategies to address three high-priority NCFs described in our previous report (Miro et al., 2022). In the earlier report, the NCFs Provide Public Safety and Supply Water were determined to be at risk of moderate disruption on a national scale by 2030 and were deemed high priority for risk mitigation investment, while the NCF Distribute Electricity was determined to potentially create cascading effects on the greatest number of other NCFs if disrupted. We identified 19 unique adaptation strategies for Supply Water and 24 unique strategies for Distribute Electricity. However, we identified only four unique strategies for Provide Public Safety, all of which are concentrated on addressing workforce shortages and have historically been challenging to implement.

FIGURE S.1

Example of How an Adaptation Strategy Interrupts the Impact Pathway for an NCF

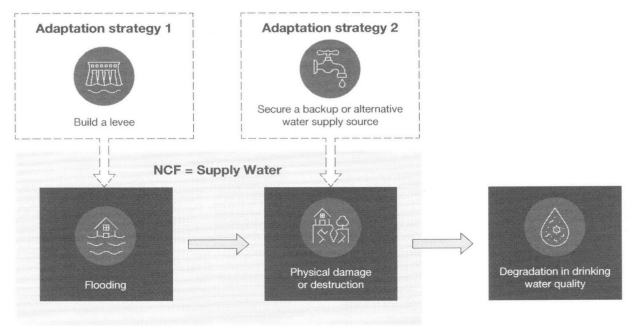

Across all 25 NCFs, we identified 254 unique adaptation strategies to mitigate risk from climate change. These strategies address a total of 179 impact pathways that we identified across the NCFs. We mapped the 254 adaptation strategies against each relevant impact pathway for each NCF. For example, "building sea walls and coastal protection structures" is an effective strategy for mitigating both coastal flooding and sea-level rise climate drivers and is applicable to nine of the 25 NCFs. Information about the number of available strategies provides owner-operators of critical infrastructure, federal agencies, and other stakeholders a starting point for deciding on investments to mitigate risk from climate change.

We identified multiple strategies to address some impact pathways. In such cases, owner-operators have more options to consider in deciding how to mitigate risk to their NCFs. Several impact pathways emerged as potentially strategy rich, with ten or more adaptation strategies identified. For example, Generate Electricity, Distribute Electricity, Transmit Electricity, Transport Cargo and Passengers by Air, Transport Cargo and Passengers by Rail, Transport Cargo and Passengers by Road, Transport Cargo and Passengers by Vessel, and Provide Housing all appear to be particularly strategy rich for mitigating physical damage. Given that Distribute Electricity has the most NCFs dependent on it, it may be particularly worthy of consideration for investments in adaptation strategies.

In contrast, we identified only one adaptation strategy each for 43 impact pathways across 12 NCFs: Produce and Provide Agricultural Products and Services, Generate Electricity, Enforce Law, Prepare for and Manage Emergencies, Provide Public Safety, Maintain Supply Chains, Manufacture Equipment, Produce Chemicals, Develop and Maintain Public Works and Services, Provide and Maintain Infrastructure, Supply Water, and Manage Wastewater. While the number of strategies identified for a single pathway does not necessarily indicate the quality of the strategy, having only one strategy could be of concern if it is not feasible to implement that option in all contexts (e.g., because of the specific location, community, cost). In addition, SMEs were asked to identify at least one adaptation strategy per impact pathway; without this guidance, the team might not have identified any strategies for some impact pathways, leaving owner-operators and other stakeholders with no feasible options to consider.

The strategies can be sorted to address specific NCFs, climate drivers, or mechanisms of concern. For example, we identified 28 strategies that apply to four or more NCFs. Such strategies as sea walls and other physical structures, wetland restoration, or vulnerability assessments can mitigate climate risk across several NCFs. Similarly, we found 14 strategies that address multiple climate drivers and 40 that address more than one impact mechanism. Strategies that address more than one NCF, climate driver, or impact mechanism may be especially worthy of further investigation and possible investment because of their potential to mitigate multiple sources of risk.

How Can the Effectiveness and Feasibility of Existing Adaptation Strategies Be Assessed?

To help owner-operators find relevant strategies, we conducted a first-order assessment of the strategies' strength of evidence, feasibility, effectiveness, and impact on cybersecurity. **We assessed the majority of strategies to have moderate impact in mitigating risk from climate change.** This suggests that multiple strategies or portfolios of strategies may be needed to sufficiently reduce risk from climate change, particularly for the NCFs at greatest risk.

In general, the strategies we identified have medium to strong evidence of their relevance and effectiveness in mitigating risk. Only 5 percent of identified strategies had weak strength of evidence; these strategies were often included to ensure that at least one strategy was identified for each impact pathway.

Strategy feasibility ratings were mixed, and only a small portion of the strategies were assessed to have an impact on cybersecurity vulnerability. While low feasibility does not imply that the climate adaptation strategy is impossible, the rating does mean that the SME found the barriers to success to be significant.

What Tools Are Available to Assist Stakeholders with Climate Adaptation?

Collectively, the adaptation strategies provide a starting point for risk mitigation planning that is aligned with the NCFs. This is just the first step in the decisionmaking process. NCF owner-operators must assess the effectiveness and feasibility of potential strategies in mitigating the specific risks to their critical infrastructure. In addition, owner-operators and other stakeholders must consider other issues and trade-offs when prioritizing, selecting, and implementing adaptation strategies. These include interdependencies between NCFs and strategies, potential co-benefits or unintended consequences, climate change uncertainties, and negative impacts related to equity.

We also identified tools provided by authoritative sources, such as government agencies, that are currently available for free and could be used to assist decisionmakers in selecting from among adaptation strategies. A vast number of decision tools of varying quality and currency are available that could be used to guide vulnerability assessments and investment decisions. We began collecting those that had the most relevance to the NCFs and identified 50 guides and tools that owner-operators could use to assess potential adaptation strategies.

Recommendations

Our findings suggest several potential next steps:

Improve the candidate set of strategies. The structured approach we used can be repeated both to identify new and emerging strategies and to improve assessments of feasibility, effectiveness, and strength of evidence for strategies already identified. Specifically, future efforts could

- focus on identifying strategies for NCFs or impact pathways for which we did not identify a large number of feasible and effective strategies to ensure that viable options exist
- gather additional, more-detailed information on strategies for NCFs we identified as having a large number of adaptation strategies, including investigating the ways in which individual strategies could be implemented in combination to produce greater risk reduction and the enablers and barriers to broad implementation
- investigate strategies with broad applicability across NCFs and provide research and technical assistance as appropriate
- investigate why existing strategies have not been implemented and/or have not been implemented at the scale required to reduce risk, including identifying enablers and barriers
- include more-generalized strategies and strategies for all stakeholders, including strategies for households and strategies that are not specific to an NCF or community vulnerabilities that could affect an NCF
- regularly review and update the strategies to reflect new technologies and emerging solutions.

Address this approach's limitations. This could include accessing a broader array of sources in addition to the open-source literature, such as proprietary or commercial sources; reviewing more-general adaptation strategies that may not have been linked to a specific NCF but that could still be useful, including additional rating dimensions beyond qualitative assessments of feasibility and effectiveness; and conducting a more granular analysis of these and other strategy characteristics. Assessment ratings, such as feasibility and effectiveness, could be built out by, for example, addressing what factors influence these characteristics and how they vary at the local level.

Provide additional decision-support tools to owner-operators in the commercial sector and state, local, tribal, and territorial governments to help them select strategies and assemble packages of strate-

gies. We identified general, high-level strategies to address risk from climate change, provided a high-level assessment of the strategies; and identified 50 tools that are currently available to help choose strategies. However, specific communities will need to consider their own local contexts, resources, and a large number of factors outside the scope of this report (for example, equity) when deciding which adaptation strategies to choose. To assist with such an effort, we hope to publicly release a searchable database and tool for the strategies we identified and the information we compiled for this project. However, a more advanced decision-support tool could help decisionmakers identify and assess these considerations.

Factor the consequences of NCF disruption into future risk assessments. CISA should consider conducting a more-complete analysis of the consequences of various levels of disruption examined to inform the prioritization of future risk mitigation activities.

Focus on developing robust communication material to help stakeholders understand climate change risk and adaptation strategies. Climate change–related risk presents a daunting number of communication challenges, including uncertainty in where, when, and how specific areas will be affected by climate change; the large number of impact pathways through which risk may manifest; the number of potential adaptation strategies to address these risks; and the trade-offs, uncertainties, interdependencies, and unintended consequences of these strategies. CISA should continue to invest in robust communication tools to convey the risk the United States faces from climate change, its potential impact, and how this risk may be reduced through adaptation strategies, such as the ones identified in this report.

Contents

APPENDIXES

Figures and Tables

Figures

Tables

CHAPTER ONE

Introduction

In January 2021, President Joseph Biden issued an executive order directing the Secretary of Homeland Security to "consider the implications of climate change in the Arctic, along our Nation's borders, and to National Critical Functions [NCFs]" (Executive Order 14008, 2021). The NCFs represent "the functions of government and the private-sector so vital to the United States that their disruption, corruption, or dysfunction would have a debilitating effect on security, national economic security, national public health or safety, or any combination thereof" (Cybersecurity and Infrastructure Security Agency [CISA], undated, p. 1). The 55 NCFs represent an evolution in CISA's approach to critical infrastructure risk management. Prior approaches have largely been sector-based and have generally focused on physical assets, such as buildings. By focusing on functions, the NCFs allow improved analysis of crosscutting risks that could have cascading effects within and across critical infrastructure sectors and could be driven by factors other than risk to physical assets.

To fulfill the objectives of the executive order, CISA asked the Homeland Security Operational Analysis Center (HSOAC), a federally funded research and development center operated by the RAND Corporation, to develop a risk management framework and to assess the risk of climate change to the NCFs. In a previous report, we developed the five-step risk management framework shown in Figure 1.1 (Miro et al., 2022). In that report, we conducted an NCF climate change risk assessment and addressed the first four steps of the framework. This report addresses the fifth step: mitigating the risk.

FIGURE 1.1
Risk Management Framework

1
Screen the NCFs
- Screen rapidly
- Obtain subject-matter expert (SME) validation

2
Identify the climate drivers
- Project the changes in the eight drivers
- Characterize regional drivers

3
Define the pathways to effects
- Identify the mechanisms of each effect
- Identify the consequences for each NCF operation

4
Assess the risks
- Assess risks for eight climate drivers in three future time periods in two climate scenarios (current and high emissions)

5
Mitigate the risks
- Develop a strategy
- Assess the strategy's feasibility and effectiveness

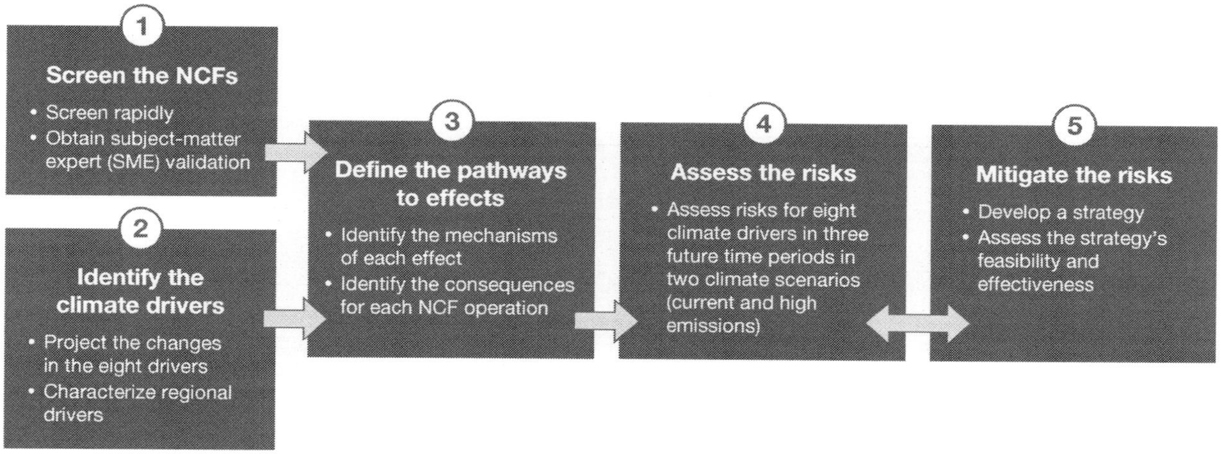

SOURCE: Miro et al., 2022.

Key Findings from the NCF Climate Change Risk Assessment

In our previous report (Miro et al., 2022), we assessed risk to NCFs from climate change at the national level using a five-point scale that is compatible with previous CISA efforts (Table 1.1). As noted in that report, "moderate disruption" on a national scale could include major historical weather events, such as Hurricane Katrina (2005), that lead to major disruption on the local or regional scale and could entail significant risks to life and safety and for economic loss.

We briefly summarize key findings from our previous report:

- **We assessed 25 NCFs to be at risk of moderate disruption on a national scale by 2100, with ten NCFs at risk of moderate disruption by 2050.**[1] The 25 NCFs at risk of moderate disruption by 2100 are the focus of this report (Table 1.2).
- **The NCFs assessed to be at greatest risk in the near term are Provide Public Safety and Supply Water,** both of which are at risk of moderate disruption by 2030.
- **Of the eight climate drivers assessed, flooding, sea-level rise, and tropical cyclones and hurricanes pose the greatest risk of disruption to the NCFs, although there are important regional distinctions in their expected impact.** Other drivers presenting risk to the NCFs are drought, extreme cold, extreme heat, severe storm systems (nontropical), and wildfires.
- **The Distribute Electricity NCF has the highest potential for cascading risk if failure should occur.** Because of the interconnected nature of U.S. infrastructure, risk to one NCF has the potential to cause cascading risk to other NCFs. Distribute Electricity has the potential to disrupt 22 of the 25 at-risk NCFs, including those supporting food production, medical care, and water supply.

Focus of This Report

This report extends our earlier work by identifying and assessing general climate adaptation strategies for the 25 NCFs identified in the NCF climate change risk assessment to be at risk of moderate disruption by 2100.[2] We were asked to address the following questions:

- What adaptation strategies are available to address climate risk to NCFs?
- How can the effectiveness and feasibility of existing adaptation strategies be assessed?
- What tools are available to assist stakeholders with climate adaptation?

---

[1] Miro et al., 2022, documents the risk analysis approach and results. The authors assessed 27 NCFs for risk of disruption from climate events (*climate drivers*) at the subfunction level (as identified by CISA). In some cases, this produced seemingly inconsistent results across NCFs as a result of variation in their underlying subfunctions. For example, the Transport Passengers on Mass Transit NCF includes a subfunction, Provide Diverse Energy Sources, but the Transport Cargo and Passengers on Rail NCF does not have a subfunction related to energy. Accordingly, our analysis of the Mass Transit subfunction includes vulnerability to energy disruptions, while the Rail subfunction does not. CISA provided the subfunctions and definitions, and it is our understanding that these may be subject to change in the future. We assessed the risk of disruption to each NCF under both a "business as usual" scenario and a high-emissions scenario, which included greater temperature increases and their effects. The analysis determined that 25 NCFs had at least one subfunction that was at moderate risk of disruption under the high-emissions scenario.

[2] *Adaptation* is the word most frequently used to describe climate strategies adopted to reduce vulnerability to the impact of climate change—in this case, the vulnerability of specific NCFs. While these are sometimes described as climate risk mitigation strategies, we use *adaptation* throughout this report to differentiate these strategies from climate mitigation strategies, which are adopted to reduce the likelihood of climate change and its effects.

TABLE 1.1

Definitions of NCF Risk Ratings

Risk Rating	Definition
0—Unknown or insufficient evidence	A driver's effect on this NCF or subfunction is unknown, or there was insufficient evidence to make a risk assessment
1—No disruption or normal operations	The NCF or subfunction is anticipated to meet all routine operational needs
2—Minimal disruption	Climate change is expected to affect the NCF or subfunction, but the function is expected to meet routine operational needs
3—Moderate disruption	The NCF or subfunction is anticipated to meet all routine operational needs in most but not all of the country
4—Major disruption	The NCF or subfunction is anticipated to be unable to meet routine operational needs in most of the country
5—Critical	The NCF or subfunction is anticipated to be unable to meet any of its routine operational needs across the country

SOURCE: Miro et al., 2022.

NOTE: Given the distributed nature of NCFs, major and critical disruptions affecting the entire nation would be expected to be extremely rare, although moderate disruptions are more likely.

TABLE 1.2

The 25 NCFs Assessed to Be at Risk of Moderate Disruption by 2100

Sector	NCF
Agriculture	Produce and Provide Agricultural Products and Services
Energy	Distribute Electricity Exploration and Extraction of Fuels Generate Electricity Transmit Electricity
Government and Social Services	Educate and Train Enforce Law Prepare for and Manage Emergencies Provide Housing Provide Medical Care Provide Public Safety
Industry	Maintain Supply Chains Manufacture Equipment Produce Chemicals Provide Insurance Services
Infrastructure	Develop and Maintain Public Works and Services Provide and Maintain Infrastructure
Transportation	Transport Cargo and Passengers by Air Transport Cargo and Passengers by Rail Transport Cargo and Passengers by Road Transport Cargo and Passengers by Vessel Transport Passengers by Mass Transit
Water and Waste Management	Manage Wastewater Supply Water Manage Hazardous Materials

We address each of these questions in turn. In addition, we discuss other factors to take into account when selecting a climate adaptation strategy.

Audience and Scope

There are several important considerations related to the potential audience of this report and the scope of the analysis we conducted.

We focus on climate adaptation strategies that are potentially relevant to owner-operators of critical infrastructure. HSOAC was asked to identify climate adaptation strategies that can be implemented by owner-operators of critical infrastructure, focusing on state, local, tribal, and territorial (SLTT) governments and private-sector stakeholders and the agencies, including CISA, that provide support. As a result, we do not focus on adaptation strategies that would primarily be implemented by other types of users, such as federal agencies or individual households. The material in this report may be of interest to government or private-sector stakeholders who may wish to use the NCF framework to glean general observations about the distribution of climate adaptation strategies across the impact pathways. The findings should also be of interest to CISA; multisector groups with a focus on critical infrastructure, such as sector coordinating councils; sector risk management agencies and other federal agencies and partners managing risk to U.S. critical functions; and policymakers, research analysts, and other stakeholders interested in addressing the impacts of climate change.

This report does not make specific recommendations regarding the value of one strategy over another; the choice of one or more strategies depends on a host of considerations that are specific to the critical infrastructure owner-operator, including location; local priorities and resources; and existing planning processes, such as capital improvement or land-use plans. Instead, we identify one or more strategies that are relevant to each of the 25 NCFs and could be worthy of further consideration by an individual owner-operator.

We frame our analysis around NCFs and, more specifically, around the impact pathways through which climate change might disrupt a particular NCF. As we will explain further in Chapter Two, climate change has the potential to disrupt NCFs in various ways. For example, a hurricane might damage physical infrastructure at a water plant, disrupting the NCF Supply Water, or a flood might increase the demand for police officers and other first responders, contributing to workforce shortages that disrupt the NCF Provide Public Safety. Given the intended audience of this report, we frame our analysis around the NCFs and the impact pathways that can disrupt them so that a stakeholder with responsibility for a vulnerable NCF can assess the availability of adaptation strategies for that NCF. In some cases, climate change might pose multiple risks to one NCF, and owner-operators will need to consider all relevant impact pathways.

As we discuss in more detail throughout the report, many adaptation strategies can potentially apply to more than one NCF and to more than one impact pathway. Thus, an alternative way to organize this report would have been to frame the report around the strategies themselves. We include some analysis framed around the strategies at some points in this report while maintaining our focus on our primary task of assessing the availability of strategies at the NCF level.

We focus on identifying general adaptation strategies and include only a first-order, high-level assessment of key characteristics of the strategies, such as feasibility and potential effectiveness. This focus was driven by several practical considerations. The purpose of this report is to provide a road map for a broad range of stakeholders and to identify where relevant strategies may be available and provide a general assessment of how feasible and effective the strategies are likely to be. At the same time, we recognize that adaptation strategies will inevitably be implemented in a specific community, by a specific set of stakeholders, in a specific context, and at a specific scale. The sheer range of strategies and potential contexts made it impossible, within project resources, to identify more-specific strategies or to conduct a more granular analysis of feasibility and effectiveness.

However, in Chapter Four, we identify freely available decision-support tools and indicate where they can be accessed to assist stakeholders and include citations for each strategy to enable users to explore strategies of interest in more detail. We also note when we were not able to identify relevant, feasible, and effective strategies for a particular NCF, suggesting that further efforts to identify relevant strategies may be needed.

Finally, in identifying adaptation strategies, we focus especially on areas that CISA has the capability to assist with, including critical infrastructure and cybersecurity. This includes a general assessment of whether implementing a potential adaptation strategy may be likely to increase or decrease cybersecurity risk. Other adaptation strategies may be available that we did not find.

Spreadsheet Tool for Selecting Climate Adaptation Strategies

The detailed results of our analyses are contained in separate Excel spreadsheet files, which also include the full list of strategies, sources describing the use of the strategies, our ratings of the strategies, and some existing tools that could be used to guide decisions on which strategy or set of strategies to pursue. We also developed a Tableau tool that facilitates exploration and characterization of the database of strategies.[3]

We were not able to include the spreadsheet tool in this report. However, we explain the contents and organization of the tool, which we hope to make available to the public at a future date. Appendix A presents information on the strategies, organized by NCF.

If and when the spreadsheet tool is made public, critical infrastructure owner-operators and others responsible for providing NCFs might use the information contained in the tool to better understand their risks from climate change and could provide a starting point for identifying strategies to consider to reduce these risks. Stakeholders have many options in terms of climate adaptation strategies to pursue, and the information provided in these materials could help stakeholders identify specific actions that hold promise for their particular needs, geographies, and capacities.

Terminology Used in This Report

Throughout this report, we use the general terminology shown in Table 1.3 to define and describe our methodology and findings.

Climate Drivers and Impact Mechanisms

As stated previously, an impact pathway consists of a unique combination of an NCF, a climate driver, and an impact mechanism. In this report, we refer to the eight climate drivers shown in Table 1.4 and the four impact mechanisms shown in Table 1.5. These are derived from our previous report (Miro et al., 2022).

---

[3] Tableau is proprietary software that allows users to create visualizations of information for analysis.

Organization of This Report

The remainder of this report is laid out as follows:

- Chapter Two describes the methods used to identify, categorize, and assess climate adaptation strategies.
- Chapter Three provides an overview of the adaptation strategies and discusses our high-level assessment of their effectiveness and feasibility.
- Chapter Four notes some considerations that critical infrastructure owner-operators should address when selecting adaptation strategies to reduce risk to NCFs.
- Chapter Five offers our conclusions and next steps.

TABLE 1.3

Terminology Used in This Report

Term	Explanation
NCFs	"The functions of government and the private sector so vital to the United States that their disruption, corruption, or dysfunction would have a debilitating effect on security, national economic security, national public health or safety, or any combination thereof" (CISA, undated, p. 1).
Climate drivers	Climate-related hazards that pose risk of disruption to NCFs. The eight climate drivers are drought, extreme cold, extreme heat, flooding, sea-level rise, severe storm systems (nontropical), tropical cyclones and hurricanes, and wildfires. These are further defined in the next chapter and shown in Table 1.4.
Impact mechanisms	Mechanisms that characterize how climate change causes a risk of disruption to an NCF, such as through physical damage or workforce shortages. While a wide range of mechanisms exists, for the purposes of this analysis, we categorize them into four groups: physical damage or disruption, input or resource constraint, workforce shortage, and demand change. These are shown in Table 1.5.
Impact pathway	The path by which a climate driver can disrupt an NCF via an impact mechanism. Each impact pathway represents a unique combination of an NCF, climate driver, and impact mechanism. The impact pathways included in this analysis are those that pose risk of disruption to an NCF.
Climate adaptation strategy	An investment or intervention that mitigates the risk posed to an NCF due to climate change. Adaptation strategies were developed for each NCF to mitigate the risks from each climate driver. Adaptation strategies were also categorized according to the impact mechanisms they address.
Adaptation strategy–impact pathway pairing	A pairing of an adaptation strategy with an impact pathway. More than one adaptation strategy may mitigate risk for an impact pathway, and more than one impact pathway may be relevant for an adaptation strategy.
25 highest-risk NCFs	The NCFs included in this report. These NCFs are those at risk of a moderate disruption that is due to climate change by 2100, as assessed in our preceding climate risk assessment report (Miro et al., 2022).
SME	An expert in an NCF by virtue of professional background.

TABLE 1.4
Climate Drivers

Climate Driver	Description
Drought	An extended period of moisture deficiency with impacts on dependent systems; can occur on seasonal to multidecadal time scales
Extreme cold	Extreme cold temperatures
Extreme heat	Extreme high temperatures and heat waves
Flooding	Riverine and flash flooding from extreme rainfall events
Sea-level rise	Coastal inundation from sea-level rise
Severe storm systems (nontropical)	Convective storm systems, extratropical cyclones, nor'easters, and associated hazards, including hail, extreme rainfall, snow, and ice
Tropical cyclones and hurricanes	Tropical cyclones and hurricanes, as well as their associated weather effects (storm surge and waves, high winds, extreme rainfall)
Wildfire	Wildfire in wildlands or at the wildland–urban interface

SOURCE: Miro et al., 2022.

NOTE: These definitions are adapted from the National Oceanic and Atmospheric Administration (NOAA) National Centers for Environmental Information Storm Events Database, compiled by the National Weather Service (National Weather Service Instruction 10-1605, 2021).

TABLE 1.5
Impact Mechanisms

Impact Mechanism	Definition
Physical damage or disruption	Physical damage to facilities or equipment necessary for the functioning of the NCF; includes cases in which a climate driver might disrupt but not necessarily damage infrastructure (e.g., high temperatures that ground airplanes)
Input or resource constraint	An interruption in the supply of inputs to the NCF, preventing the NCF from functioning; these could be raw materials or goods or services produced by others, including those produced or delivered by other NCFs, such as water, power, and communication
Workforce shortage	A shortage of workers needed to operate the NCF; includes conditions in which the climate driver makes such work unpleasant or dangerous
Demand change	Changes in the demand for the NCF, resulting in the NCF not being able to fully meet the demand; includes an increase in demand that goes beyond what the NCF can supply, a decrease in demand that affects the viability or efficiency of the NCF, and volatile fluctuations encompassing both increases and decreases

SOURCE: Miro et al., 2022.

CHAPTER TWO

Methodology

In this chapter, we describe the methodology we used to identify adaptation strategies that could reduce the risks of climate change impacts for specific NCFs. First, we discuss the impact pathways that are central to our methodology and define key inputs to these pathways. Next, we describe the steps used to identify the climate adaptation strategies. Finally, we describe several limitations to our approach.

Framework for Understanding How Climate Adaptation Strategies Address Risk to NCFs

As discussed in the previous chapter, we have framed our analysis of adaptation strategies around NCFs. This means that we are interested in understanding how specific adaptation strategies can address risk from climate change to an NCF. To explore this issue, we focus on *impact pathways*. Each impact pathway represents a unique combination of an NCF, climate driver (e.g., flooding, hurricane) and impact mechanism (e.g., physical damage, increase in demand).[1]

Consider the example of the impact pathway shown in Figure 2.1. Without any adaptation strategies, the climate driver *flooding* contaminates a drinking water source through the impact mechanism *physical damage or destruction*, leading to the consequence *degradation in drinking water quality*. This process is described in greater detail in Miro et al., 2022.

A climate adaptation strategy attempts to interrupt the impact pathway for the NCF in one of two ways: either by reducing or eliminating the likelihood that a climate driver will disrupt an NCF or by reducing or eliminating the potential consequences of a climate driver on an NCF. For example, Figure 2.2 shows the addition of two potential adaptation strategies to interrupt the flow of the impact pathway. One adaptation strategy would be to *build a flood barrier*, such as a levee, to prevent a flood from reaching a drinking water treatment plant or its supply sources to prevent the physical damage. Another adaptation strategy would *secure a backup or alternative water supply source* to reduce a potential disruption in operations due to a flood.

Additionally, the same adaptation strategy may reduce the risk climate change poses across multiple impact pathways. The strategy *lengthening runways* can reduce the risk of disruption from extreme heat for both the Manage Supply Chain NCF and the Transport Cargo and Passengers by Air NCF. Strategies of this type, which benefit multiple NCFs, may be particularly promising. These strategies reduce risk to multiple NCFs, and, because a single strategy may have multiple impacts, more stakeholders may be available to support or even fund such a strategy.

---

[1]   Our prior report included a consideration of the consequences within an impact pathway. The current report focuses on the avoiding the consequences through adaptation strategies and focuses on the driver, mechanism, and NCF combination that forms an impact pathway.

FIGURE 2.1

Example Impact Pathway for an NCF

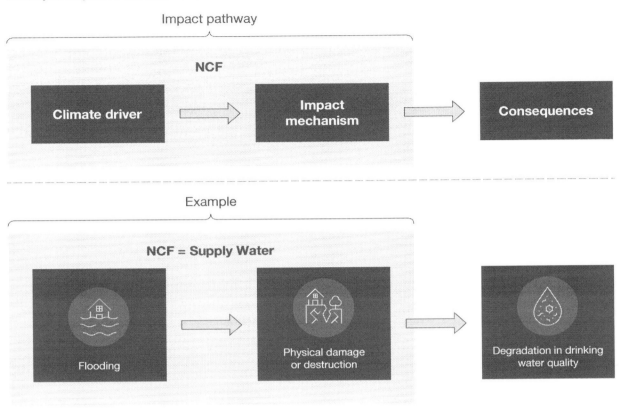

SOURCE: Miro et al., 2022.

Climate Drivers and Impact Mechanisms

As stated previously, an impact pathway consists of a unique combination of an NCF, a climate driver, and an impact mechanism. In this report, we refer to the eight climate drivers shown earlier in Table 1.4 and the four impact mechanisms shown in Table 1.5. These are derived from our previous report (Miro et al., 2022).

Process for Identifying Candidate Adaptation Strategies

To identify candidate adaptation strategies, we first sought to identify one or more climate adaptation strategies for each impact pathway. As discussed in Chapter One, we focused on identifying potential adaptation strategies that critical infrastructure owner-operators could implement, focusing on SLTT governments and private-sector stakeholders seeking to address climate risk to the NCFs.

For our analysis, we followed the four-step process illustrated in Figure 2.3. We describe each of these steps in the following subsections.

Step 1. Identify Risks to Mitigate

First, we used the risk assessment framework shown in Figure 1.1 to identify where specific adaptation strategies were most likely to be needed. We identified impact pathways associated with the 25 highest-risk NCFs (that is, those we assessed to be at risk of at least moderate disruption from climate change by 2100 in the

FIGURE 2.2

Example of How an Adaptation Strategy Interrupts the Flow of the Impact Pathway for an NCF

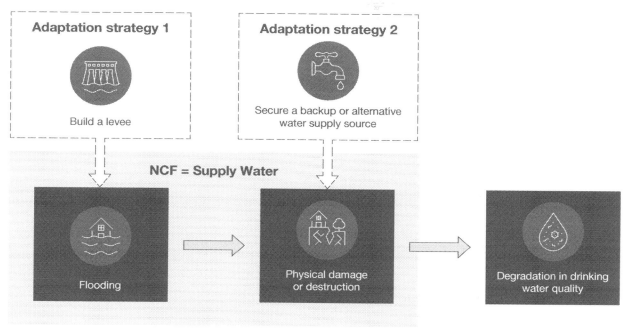

FIGURE 2.3

Adaptation Strategy Identification Method

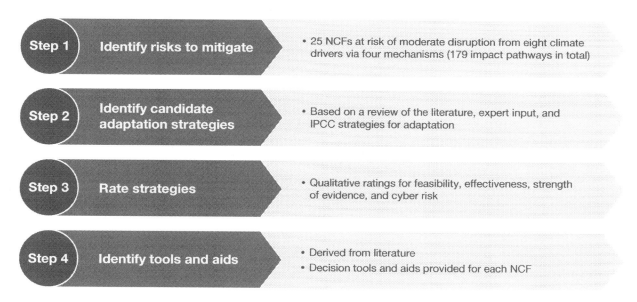

NOTE: IPCC = Intergovernmental Panel on Climate Change.

previous study). Not all NCFs were at risk of moderate disruption from all eight climate drivers for all four mechanisms. We identified a total of 179 impact pathways across the 25 NCFs, with each NCF associated with multiple pathways.[2]

---

[2] With 25 NCFs, eight climate drivers, and four impact mechanisms, there were a total of 800 potential impact pathways.

Step 2. Identify Climate Adaptation Strategies

Next, we used a multistep process to identify adaptation strategies to address each of the 179 impact pathways from Step 1, drawing on previously published research and subject-matter expertise to identify frequently cited or well-regarded adaptation strategies. It was imperative to develop an approach that could address the wide range of impact pathways and that considered and captured a comprehensive set of climate adaptation strategies at a level that provided enough information to allow stakeholders to have a basic understanding of the strategy to help them determine whether to investigate it further. The large number of impact pathways and the large number and wide range of potential adaptation strategies limited opportunities for in-depth analysis into any one strategy.

As indicated, first we considered various typologies to frame and guide our planned adaptation strategy identification and ultimately selected the Fifth IPCC climate adaptation framework categories. The IPCC is a United Nations organization that is the international authority on climate science. The IPCC provides comprehensive scientific information on the status of climate change, including the physical science of climate change, the vulnerabilities and effects of climate change, and climate change mitigation and adaptation approaches. It publishes periodic reports on the status of climate change and on special topics and methodological approaches. Using an IPCC-developed framework (shown in Table 2.1) provides several advantages:

- It links the current analysis to internationally accepted and researched categories.[3]
- It provides a benchmark to ensure that the current analysis is comprehensive and does not omit key adaptation strategies.
- It shows how the adaptation strategies we identify fit within the full range or a portfolio of what can be done across all sectors of society and all actors.

The IPCC categories include high-level strategies that may be relevant to a given impact pathway (shown in Table 2.1). As the first step in our process to identify climate adaptation strategies, we asked SMEs within HSOAC to perform an initial screening to identify how the IPCC subcategories mapped to the 179 impact pathways. We assigned at least one researcher with subject matter expertise to each NCF. These individuals were SMEs in their NCFs by virtue of their professional backgrounds and, in many cases, had previously worked with their assigned NCFs on a prior project that HSOAC conducted for CISA to assess risk to the NCFs from COVID-19 (Lauland et al., 2022). A list of SMEs appears in Appendix C.

Second, guided by the IPCC subcategories that were identified as relevant for each NCF, these SMEs identified a comprehensive list of climate adaptation strategy options for each of their respective NCFs. SMEs relied on their own knowledge of authoritative documents and also performed Google searches using appropriate terms to describe their NCF, the relevant climate drivers, and "risk mitigation" or "climate adaptation" to identify the strategies. SMEs were asked to identify at least one adaptation strategy for each of the impact pathways. We employed this approach to attempt to provide at least a starting point for each impact pathway; however, as we discuss later in the report, impact pathways for which SMEs could only identify one climate adaptation strategy may suggest areas that lack viable strategies and in which further efforts to develop adaptation strategies are particularly needed. SMEs were directed to focus on identifying strategies that are frequently discussed in practitioner and academic literatures for a specific NCF and that would most likely

---

[3] The IPCC has published a revised framework for considering adaptation options in its Sixth Assessment Report. We chose the adaptation framework from the Fifth Assessment Report for two reasons. First, the Sixth Assessment was not available at the time we were developing our research approach. Second, given our focus on risk to the NCFs (as opposed to communities or populations), the structure of the Fifth Assessment provided a useful typology for benchmarking the suite of strategies our SMEs identified. As is to be expected, the science is evolving, and the Sixth Assessment has reframed these categories to align more directly with the types of climate risks.

TABLE 2.1
IPCC Climate Adaptation Framework

Category	Subcategory	Examples
Structural/physical	Engineered and built environment	Sea walls and other coastal protection structures Flood levees and culverts Water storage and pumps Adjusting power plants and electricity grids
	Technological	Water-saving technologies Hazard mapping and monitoring technology Renewable energy
	Ecosystem-based	Ecological restoration of wetlands Floodplain conservation Adaptive land-use management
	Services	Municipal services Essential public health services
Social	Educational	Awareness raising Extension services Community surveys
	Informational	Hazard and vulnerability mapping Early warning and response Systematic monitoring
	Behavioral	Individual planning and preparedness Retreat Crop switching
Institutional	Economic	Financial incentives—taxes or subsidies Insurance design Payments for ecosystem services
	Laws and regulations	Land zoning laws Building standards and codes Protected areas or easements
	Government policies and programs	National and regional adaptation plans SLTT preparedness Municipal water management programs Landscape and watershed management

SOURCE: Noble et al., 2014.

be implemented by owner-operators in either SLTTs or the private sector, which represent the core audience CISA identified for this report.

We conducted a final check by crosswalking the identified climate adaptation strategies for each NCF against the IPCC Climate Adaptation Framework subcategories to check for completeness, recognizing that we were focused on the subcategories of adaptation strategies most relevant to SLTTs and the private sector. These results were independently reviewed and validated by other researchers who also had subject-matter expertise in the NCFs. We include a list in Appendix C as well.

Step 3. Rate Strategies on Effectiveness, Feasibility, Cybersecurity Vulnerability, and Strength of Evidence

After identifying adaptation strategies for each impact pathway, we assessed each adaptation strategy using a set of standardized criteria to provide additional information to potential users. We identified four high-level criteria based on discussions with CISA personnel regarding their priorities and the types of information that they believed would be of greatest use to stakeholders in making their own assessments about which strategy to employ. We did not ask stakeholders directly what types of information would be most useful to them as it was beyond the scope of this study. We assessed adaptation strategies along four dimensions: their likely effectiveness, their likely feasibility, whether implementing the strategy increased or decreased cyber vulnerabilities to the NCF, and the strength of evidence supporting our assessment of the effectiveness of the strategy.

We assessed the adaptation strategies at a high level and considered each strategy in its most general form. Adaptation strategies will inevitably be implemented in a specific community, by a specific set of stakeholders, and in a specific context. As a result, it is difficult to assess several of the characteristics we selected in anything other than general terms. For example, the actual feasibility and effectiveness of an individual adaptation strategy, such as *Build seawalls and coastal protection structures*, may vary greatly between a major metropolitan area, such as New York City, and a smaller coastal town in another region of the country. Similarly, an entity evaluating which adaptation strategy to implement to mitigate risk to an NCF as a result of climate change may have already tried to implement a given strategy. Accordingly, we roughly sorted adaptation strategies into general categories to provide these potential strategy users with a high-level assessment.

Entities considering implementing a strategy will need to make their own assessments, which may include such considerations as whether a given adaptation strategy has already been attempted. In our discussion of Step 4, we will offer tools to assist potential users with conducting their own more granular assessment. The definitions of our criteria and how we assessed the adaptation strategies against them are provided in the following subsections. We discuss considerations for potential adaptation strategy selection further in Chapter Four.

Effectiveness

Our rating of the potential effectiveness of an adaptation strategy was based on an SME assessment of the extent to which the strategy would be expected to reduce risk that each relevant impact pathway to the NCF would pose if the adaptation strategy were implemented in isolation. Each adaptation strategy was assigned one of the following ratings:

- *No impact*—If this were the only adaptation strategy implemented, it would have no impact on risk.
- *Minor impact*—If this were the only strategy implemented, it would have a minor impact on reducing risk.
- *Moderate impact*—If this were the only strategy implemented, it would have a moderate impact on reducing risk.
- *Major impact*—If this were the only strategy implemented, it would have a major impact on reducing risk.

Given the variety of ways and contexts in which an adaptation strategy could be implemented, we asked SMEs to determine what a minor, moderate, or major impact meant in the context of the impact pathway they were assessing the strategy against. During the analysis process, a full-team discussion was held to calibrate how the SMEs were applying the rating rubric to their NCFs to ensure as much consistency in approach as practicable. The actual effectiveness of an adaptation strategy implemented in a real-world context depends on too many external factors, including scale and geography, to make a more specific statement about impact. Accordingly, these ratings are inherently subjective and are best characterized as a high-level, general assessment that is useful for sorting strategies into broad categories.

We asked SMEs to base their assessments on what impact the adaptation strategy, if implemented in isolation, would have on reducing risk. When assessing strategy effectiveness, SMEs did not consider the aggregate risk reduction from packages of adaptation strategies or the upstream or downstream impacts on other NCFs at this stage of the analysis. Adaptation strategies rated as "no impact" for a particular impact pathway were screened out and are not included in the final dataset; we included this rating level because strategies could be assigned different effectiveness ratings for different impact pathways, and, as described earlier, we asked SMEs to consider and rate strategies that were identified by other SMEs. We provide descriptive statistics on the number of strategies assessed at each level of effectiveness in Chapter Three.

Feasibility

We asked SMEs to consider multiple factors in assessing the potential feasibility of an adaptation strategy. Each strategy was assigned one of the following feasibility ratings:

- *High*—The strategy is relatively easy to implement; its costs are not prohibitively high; it can be implemented by one sector or unit of government; it does not require large amounts of land or skilled labor; and the technologies are widely available.
- *Medium*—Some barriers to implementation exist: Costs are high; the many actors need to cooperate; and/or there is intensive need for land, skilled labor, or technologies not currently in widespread use.
- *Low*—The strategy is difficult to implement for multiple reasons, such as high cost; the many actors need to cooperate; there is an intensive need for land, skilled labor, or technologies not currently in widespread use.

We included multiple considerations because the feasibility of implementing a strategy is likely to depend on more than one factor (the IPCC notes further that feasibility is ultimately both context- and time-dependent).[4] We included the following as considerations: costs; whether multiple parties needed to cooperate to implement the strategy; and whether implementing the strategy produced an intensive need for land, skilled labor, or technologies not currently in widespread use.

We also considered the compounding impact on feasibility if more than one of these factors was likely to be present. If an adaptation strategy is expensive or requires cooperation between several actors, either of these factors alone could reduce the feasibility of implementing the strategy; however, feasibility would be reduced even more if both factors were present. For example, we assessed that the strategy *Maintain vegetation to reduce risks of debris and erosion* would have a high feasibility rating for roadway flooding because it is not extremely costly and can be implemented by the agency responsible for the right-of-way and if the technologies are widely available. On the other hand, *Move roads at risk* was assessed as having low feasibility because it includes high construction costs and identifying new rights-of-way, which may involve multiple government agencies and, possibly, private landowners, even though the technologies are widely available.

We assessed the feasibility of each adaptation strategy in isolation. However, feasibility may differ if a strategy were to be implemented as part of a portfolio of actions. As with effectiveness, our assessment of feasibility is inherently subjective. In real-world implementations, the factors we considered will vary widely based on local context. We determined that it was not possible to develop robust, meaningful quantitative standards for the factors we considered within project resources and given the potential range of implementation scenarios. We provide descriptive statistics on the number of adaptation strategies assessed at each level of feasibility in Chapter Three.

Cyberthreats

We assessed each adaptation strategy's potential to change the level of cybersecurity risk to an NCF based on whether the strategy would make the NCF more or less vulnerable to cyber-related disruptions. There may be instances in which the introduction of new technologies to use resources more efficiently may increase cyber

---

[4] Our definition is consistent with the IPCC description of implementation feasibility, which is influenced by economic, ecological, technological, geophysical, sociocultural, and institutional factors within a given context (Pörtner et al., 2022).

vulnerabilities by expanding the potential attack surface (for example, introducing new technology to reduce the generation of wastewater in industrial processes may increase opportunities for introducing cyberattacks). We assessed that expanded networking and monitoring of processes, while potentially beneficial from a climate risk mitigation point of view, would increase the number of potentially exploitable cyber vulnerabilities. This analysis does not make any assumptions about the degree to which newer devices or systems are more or less secure from cyberthreats.

Another example of potentially greater cyber vulnerability involves the adaptation of more autonomous systems. For example, networked transshipment at ports could reduce wait times and more efficiently offload and transfer cargo but could also create additional opportunities for cyberthreats through integrity or availability attacks on networked systems. More-distributed, independent systems, such as microgrids, could also be subject to cyberattack, but we assessed that the overall impact to an NCF could be reduced because these systems would have fewer points of connection to other like systems. In these cases, cyber contagion—the spread of a virus across multiple systems—could be reduced. At the same time, these microgrids individually could have expanded attack surfaces because of the increased sensors and localized networking to support more efficient management of system resources. Each strategy was assigned one of the following cyber vulnerability ratings:

- *Adds*—The strategy increases the NCF's vulnerability to cyberthreats.
- *Does not change*—The strategy does not change the vulnerability of the NCF to cyberthreats.
- *Reduces*—The strategy lowers the NCF's vulnerability to cyberthreats.

We provide descriptive statistics on our assessment of how strategies affected cyber vulnerabilities in Chapter Three.

Strength of Evidence

Our final assessment category was the strength of evidence for each adaptation strategy. We assessed each climate adaptation strategy based on the relative strength of evidence for it in the existing literature and the availability of authoritative sources supporting the strategy. SMEs augmented their own knowledge of authoritative documents with Google searches using appropriate terms to describe their NCFs, the relevant climate drivers, and the phrases "risk mitigation" or "climate adaptation." SMEs exercised their own judgment about what constituted authoritative sources for the field represented by the NCF. The strength-of-evidence rating has implications for other characteristics we rated. It refers to the strength of evidence supporting the use of the strategy to mitigate risk to the specific NCF. We considered the effectiveness rating an assessment of the extent to which the strategy, if implemented, would mitigate risk, and the strength-of-evidence rating as a measure of the confidence in that rating. Each strategy was assigned one of the following strength-of-evidence ratings:

- *Weak*—Few to no existing studies or authoritative sources discuss the use of this strategy for mitigating risk to this NCF.
- *Medium*—A few, well-cited studies or authoritative sources discuss the use of this strategy for mitigating risk to this NCF.
- *Strong*—A wide body of work and authoritative sources discusses the use of this strategy for mitigating risk to this NCF.

We provide descriptive statistics on the number of strategies assessed at each level of strength of evidence in Chapter Three of this report.

Step 4. Identify Climate Adaptation Tools and Aids

Finally, at CISA's request, we also identified existing tools and other aids (referred to hereafter as *tools*) that stakeholders could use to inform their decisionmaking when assessing adaptation strategies to reduce risk from climate change to an NCF or subfunction. To develop an initial list, we searched for tools and other aids that had been developed by federal agencies, such as the U.S. Environmental Protection Agency (EPA), in existing databases of potentially relevant resources, such as the U.S. Climate Resilience Toolkit webpage (U.S. Climate Resilience Toolkit, undated) and the Adaptation Clearinghouse webpage (Adaptation Clearinghouse, 2022). We then reviewed the initial list to identify the subset of tools that were specifically relevant to owner-operator investment decisionmaking for climate adaptation strategies.

SMEs for each NCF reviewed this prioritized list and assessed whether each tool was relevant to an individual NCF. At this stage, SMEs were also able to add relevant tools that they identified but that were not on the original list; tools identified in this manner were then assessed for relevance to other NCFs by their respective SMEs. We compiled the information on the resulting final set of tools into a spreadsheet that includes the name of each tool, its relevance to a given NCF, the developer of the tool, and a link to the tool itself. However, the resulting list is not a comprehensive view of all available tools for informing selection of adaptation strategies. We excluded tools that

- **Are location-specific.** Many states and local communities have developed tools focused on climate change risk adaptation that are tailored to their specific needs and use local data. For instance, the state of New Jersey developed a toolkit to support municipalities in developing climate change–resilient strategies at a community level (U.S. Climate Resilience Toolkit, undated). While these are essential tools for decisionmaking, our focus was on national-level tools that are readily accessible for early-stage screening.
- **Focus on "upstream" analysis,** such as climate change risk assessment and climate change exposure
- **Are proprietary or commercial and available only through membership or purchase.** For example, the American Water Works Association has published a manual on adaptive management strategies for water utilities, which is available to purchase through the website (American Water Works Association, 2021).

Compiling the Strategies into a Database and Tool

Appendix A presents the full list of unique adaptation strategies by NCF. At the National Risk Management Center's request, we compiled the candidate strategies, the impact pathways they mitigate, our assessments on the four criteria described above, supporting citations, and associated tools into a spreadsheet matrix that was delivered to CISA on March 4, 2022. We also provided a Tableau tool that facilitates exploration and characterization of the database.

As mentioned previously, the candidate adaptation strategies themselves are generalized and address the risk posed to an NCF from a climate driver via an impact mechanism. We gathered standard information for each adaptation strategy that includes the name and description of the strategy (or an exemplar); the IPCC adaptation category it maps to most closely; the source(s) for the strategy; and a qualitative assessment of the strength of evidence for the strategy, its feasibility, and its effectiveness. This information is included in the database and tool. We also flagged whether the strategy has the potential to increase or decrease cybersecurity vulnerabilities. To illustrate the type of information gathered, Table 2.2 is an example for one candidate strategy for Distribute Electricity.

Limitations of the Approach

Our approach has several limitations that have implications for interpreting our findings and for future efforts. Here, we discuss the primary limitations we identified.

First, how we identified climate adaptation strategies had limitations. We relied on open-source literature and readily available practitioner literature to identify adaptation strategies. We did not review proprietary or commercial tools and other materials that could include information on additional strategies or more-definitive information on their benefits or feasibility.[5] Our review was also limited to strategies that explicitly mention climate risk and at least one individual NCF. There may be other general risk mitigation approaches or strategies that may reduce climate risks, even though climate change or a specific NCF is not identified explicitly. We also focused on gathering strategies with strong evidence first and, as a result, may have under-counted or ignored very effective but novel strategies.

We identified strategies using an iterative process in which SMEs in each NCF identified strategies for their respective NCFs. We created a common pool of strategies so that strategies identified for one NCF could be used for other NCFs. We had a second team of independent SMEs review the resulting set of strategies on an NCF-by-NCF basis. While these measures improved consistency across analysts, there may still be variability at the analyst level regarding how strategies were identified. We have attempted to provide a transparent methodology and comprehensive set of results so that future researchers can review and build on our findings.

We were also limited in how many characteristics of each strategy we were able to assess and in how we assessed them. We assessed strategies along four dimensions: effectiveness, feasibility, strength of evidence, and whether they added or reduced cyber vulnerabilities. While these characteristics are important, we did not assess many other potentially pertinent dimensions. In addition, we did not conduct a comprehensive analysis of the characteristics we did assess. The large number of NCF–climate driver–mechanism combinations for which strategies needed to be identified and the resulting number of strategies made it impossible to conduct a comprehensive analysis of costs and the many other factors related to feasibility within the time and resources we had available. In addition, these dimensions are likely to be highly specific to a particular community or project. To address this, we identified tools that could help entities considering a strategy

TABLE 2.2
Sample Climate Adaptation Strategy

NCF	Distribute Electricity
Climate driver	Flooding
Mechanism	Physical damage or disruptions
IPCC category	Technological
Exemplar strategy	Deploying distributed generation: distributed photovoltaic (PV) cells, microgrids, and minigrids
Citation	Gholami, Aminifar, and Shahidehpour, 2016
Strength-of-evidence rating	Strong
Effectiveness rating	Moderate impact
Feasibility rating	Medium
Cyber vulnerability rating	Increases vulnerabilities to cyberthreats

---

[5] For example, American Water Works Association, 2018; American Water Works Association, 2019, and American Water Works Association, 2021.

assess these and other factors in more depth. Future analysis could include dimensions we did not assess or a more comprehensive analysis of those we did. Some of the characteristics we identified and considered are time frame, required resources, degree of cooperation between actors needed, and more-specific dimensions of cost (such as up-front versus ongoing costs and which actors incur the costs).

A final limitation is that we considered strategies in isolation. Investments in climate adaptation strategies for one NCF–climate driver–mechanism combination may also mitigate risks to others, but we did not capture these spillover benefits in our rating system. Similarly, we did not address potential unintended consequences of implementing any given strategy. Climate adaptation strategies are also generally more effective when used as part of a more comprehensive portfolio of actions. We identified these limitations and other relevant considerations when selecting between strategies, but future efforts could provide users with decision support tools that would allow them to consider co-benefits, unintended consequences, and evaluate strategies as packages rather than in isolation.

To build on the current work, future analysis could address a broader array of sources, including proprietary ones; review more-general adaptation strategies; and review additional rating dimensions. However, it would be analytically challenging to account for all potential spillover effects and the synergies between multiple strategies, given that the ratings for many adaptation strategies would differ substantially depending on the circumstances of their implementation.

CHAPTER THREE

Overview of Climate Adaptation Strategies

In this chapter, we present the results of our effort to identify climate adaptation strategies for the 25 at-risk NCFs, i.e., those assessed to be at risk of at least moderate disruption from climate change by 2100. Two research questions are addressed in this chapter:

- What adaptation strategies are available to address climate risk to NCFs?
- How can the effectiveness and feasibility of existing adaptation strategies be assessed?

Our third research question, regarding tools that are available to assist stakeholders with climate adaptation, is covered in the next chapter.

To respond to these questions, we provide summary information on the strategies and their ratings for all 25 NCFs at moderate risk in this chapter. The strategies can be sorted to address specific NCFs, climate drivers, or mechanisms of concern. They can be sorted by strength of evidence, feasibility, effectiveness, and cybersecurity concern. Collectively, they provide a starting point for risk mitigation planning that is aligned with the NCFs.

We begin with three NCFs that are of particular interest. In our prior report (Miro et al., 2022), two NCFs, Provide Public Safety and Supply Water, were determined to be at risk of moderate disruption *by 2030* and were deemed high priorities for risk mitigation investment. Another NCF, Distribute Electricity, was determined to potentially create cascading effects on the greatest number of other NCFs if disrupted. We highlight our adaptation strategy findings for these three priority NCFs.

Next, we provide an overview of the adaptation strategies identified for the full range of 179 impact pathways. We also identify which adaptation strategies address risks across multiple impact pathways. These are strategies that may merit additional consideration, given their potential to provide additional co-benefits to other NCFs.

Finally, we provide an overview of our assessment of the effectiveness and feasibility of the strategies we identified. Appendix A provides the full set of adaptation strategies we identified and their high-level feasibility, effectiveness, and strength-of-evidence ratings.

Understanding the Results

As noted above, the climate adaptation strategies can be sorted to address specific NCFs, climate drivers, or mechanisms of concern. To display the results of our analysis, we have developed a number of tables that have several common features. Figure 3.1 shows an example row from a table to provide an example of how to read the tables. In Figure 3.1, the example row represents the NCF Supply Water, while the columns refer to four impact mechanisms, and the colored squares indicate the climate drivers as shown in the legend. Thus, each colored cell represents one impact pathway. The numbers inside the colored squares indicate the number of adaptation strategies identified for that particular pathway.

FIGURE 3.1

Example Table Row for Impact Pathways Related to the NCF Supply Water

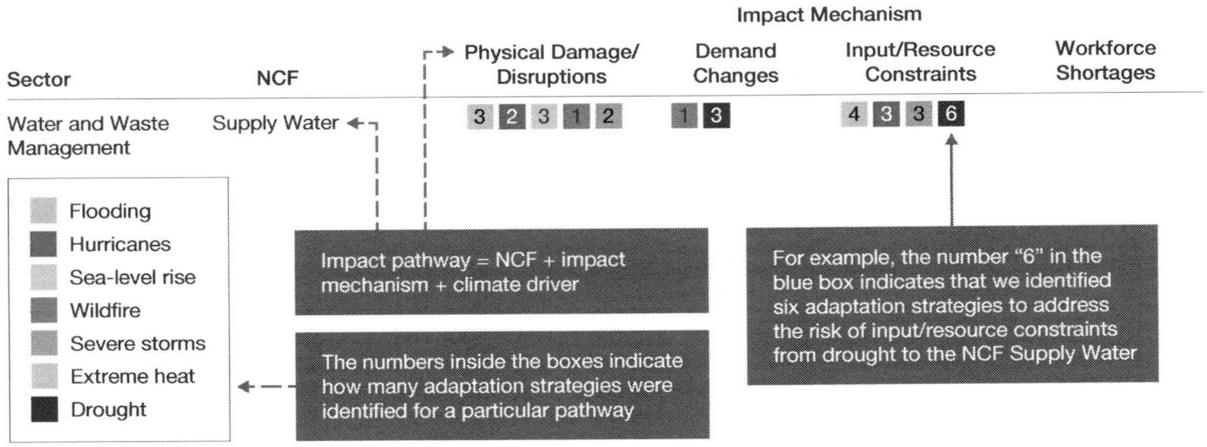

NOTE: The same strategy may apply to more than one impact pathway. For example, the two strategies identified to address risk of physical damage from hurricanes are the same two strategies identified to address risk of physical damage from severe storm systems.

For example, the green box under "Physical Damage/Disruptions" indicates that sea-level rise (green) creates a risk of disrupting the NCF Supply Water (through physical damage). The "3" inside the green box indicates that we identified three adaptation strategies to potentially address this risk. If no boxes appear under a particular impact mechanism column (as in the Workforce Shortages column in the figure), that impact pathway was not assessed to be at moderate-level risk from climate change.

Throughout this chapter, we will use variations of the spreadsheet depicted in this figure, in some cases focusing on one NCF and in other cases showing a matrix of multiple NCFs.

Strategies Identified to Address the Three Highest-Priority NCFs

This section presents what we learned about strategies to address the three highest-priority NCFs: Supply Water, Provide Public Safety, and Distribute Electricity (Miro et al., 2022).

Supply Water

Supply Water is closely connected to climate. In particular, the supply of raw water varies with climate conditions, such as drought. In addition, infrastructure and operational systems are vulnerable to extreme events, such as drought, flooding, and wildfires, which are projected to increase in frequency and severity in the future. These climate drivers pose a risk of a moderate disruption for this NCF, which could mean significant regional disruption or failure, affecting the provision of water for residential, agricultural, or industrial use. As a result, there could be significant public health and economic consequences.

We identified 19 unique adaptation strategies that address the 30 impact pathways that pose a risk of moderate disruption to the Supply Water NCF (Table 3.1). Note that this table focuses on a single NCF and lists the adaptation strategies in the left column, while the colored boxes indicate which climate drivers and which impact mechanisms are addressed by each strategy listed.

Many of these strategies are focused either on limiting physical damage or on input or resource constraints across multiple climate drivers. The adaptation strategies identified for this NCF are varied and include technological, engineering, ecological, and educational approaches (see Table 2.1 for the IPCC categories).

TABLE 3.1

Adaptation Strategies Addressing Supply Water Impact Pathways

Strategy	Impact Mechanism		
	Demand Changes	Input or Resource Constraints	Physical Damage or Disruption
Acquire and manage ecosystems and other land conservation approaches to benefit water utilities and water supply		■ ■ ■	
Build flood barriers to protect infrastructure or relocate facilities to higher elevations			■ ■ ■ ■
Build infrastructure needed for aquifer storage and recovery and increased water storage capacity		■	
Diversify options for water supply and expand current sources, including facilities to recycle water		■	
Finance and facilitate systems to recycle water		■	
Implement natural and green infrastructure on site and in municipalities		■ ■ ■	
Implement policies and procedures for post-fire repairs			■
Implement policies and procedures for post-flood repairs			■
Implement saltwater intrusion barriers and aquifer recharge			■
Implement water conservation programs to reduce water demand	■		
Implement watershed management		■ ■ ■ ■	
Improve modeling for electricity and agriculture/irrigation water demand	■		
Install low-head dam for saltwater wedge and freshwater pool separation			■
Integrate flood management and modeling into land-use planning		■	
Plan and establish alternative or on-site power supply			■ ■ ■
Practice conjunctive use		■	
Practice water conservation and demand management by implementing public outreach efforts	■		
Update drought contingency plans		■	
Update fire models and practice fire management plans	■		

■ Drought ■ Hurricanes ■ Severe storms
■ Flooding ■ Sea-level rise ■ Wildfire

We assessed the effectiveness of most of these adaptation strategies, when implemented in isolation, to be moderate.[1] We found evidence for the effectiveness of these strategies to be strong across the board but assessed feasibility as moderate in many cases. More specifically, implementation of some strategies may be challenging in areas where land values are high (affecting the cost of land conservation purchases or easements) or where new infrastructure requirements are high (e.g., water recycling infrastructure). The preva-

---

[1] Appendix A includes our assessment ratings and citations for each of the strategies discussed in this section and throughout the report.

lence of the impact mechanisms physical damage or disruption and input or resource constraints across the impact pathways for this NCF suggests that several complementary strategies would need to be implemented to achieve major reductions in climate change risk.

Provide Public Safety

Provide Public Safety was the second NCF identified in our previous report (Miro et al., 2022) that was assessed to be at risk of moderate disruption by 2030. A number of the climate drivers—tropical cyclones and hurricanes, severe storm systems, and floods, all exacerbated by sea-level rise, and wildfires—may cause widespread damage to homes, businesses, and other facilities. Cold-weather events can cause injuries and deaths, and extreme-heat events can increase heat-related illnesses. High energy usage driven by heat waves can cause blackouts, and some studies have shown that hotter weather is correlated with an increase in crime and terror attacks (Abbott, 2008; Goin, Rudolph, and Ahern, 2017). Together, the effects of climate change are expected to increase the need for law enforcement to maintain order, fire services to suppress fires and provide rescue, and emergency medical services to provide prehospital care. Service providers will also need to help conduct evacuations and support displaced people. During large disasters, emergency management personnel will need to coordinate with public health, while restoration of damaged utilities will require support from public works (Calma, 2021).

We identified four unique strategies for Provide Public Safety (shown in Table 3.2). These adaptation strategies had broad applicability to multiple climate drivers and to two mechanisms, covering a total of 20 impact pathways. The strategies generally focus on protecting physical assets and meeting demand changes, which is notable because changes in demand for the NCF in response to the climate drivers may not be well understood. While we assessed these strategies as feasible, financial constraints could preclude implementing them in many communities, particularly in regard to increasing staffing. In addition, this strategy and building a volunteer reserve corps are dependent on labor supply. While we rated these strategies as likely to succeed if implemented, public safety agencies have historically struggled to fill vacancies, and these issues have been exacerbated over the past several years. In particular, firefighting capacity, a key component of public safety, is already experiencing stress and disruption. This suggests that the ability of public safety agencies to implement these strategies remains in question. The non–workforce-oriented strategies, developing and exercising mutual aid agreements and relocation plans, may already have been attempted or may already be in place in many communities and may not be possible to scale up further.

TABLE 3.2

Adaptation Strategies Addressing Provide Public Safety Impact Pathways

Strategy	Impact Mechanism	
	Demand Changes	Physical Damage or Disruption
Build a volunteer reserve corps	▢ ■ ▢ ▢ ▦	
Develop and exercise mutual aid agreements		▢ ■ ▢ ▢ ▦
Develop and exercise plans for relocating facilities	▢ ■ ▢ ▢ ▦	
Increase staffing	▢ ■ ▢ ▢ ▦	

▢ Flooding   ▢ Sea-level rise   ▦ Wildfire

■ Hurricanes   ▢ Severe storms

While these adaptation strategies directly address the climate drivers and mechanisms identified in this work, many strategies could indirectly affect the impact pathway, such as increasing household preparedness or reducing community vulnerabilities derived from inequities or weak economic or social systems that would be worth pursuing to alleviate the risk to this NCF. Given the significant risks climate change poses for this NCF, further work to identify additional adaptation strategies for this NCF could be helpful. An additional potential interpretation is that, lacking other viable strategies, heavy investments should be made in those that are available and still have capacity to be scaled up (for example, funding to support recruiting and retaining additional public safety staff).

Distribute Electricity

We found in our previous report (Miro et al., 2022) that failure of the Distribute Electricity NCF has the greatest potential to cause cascading risk to other NCFs that depend on it to ensure normal operations. For example, a substation failure can cause power outages throughout the entire power distribution network, resulting in power outages for all interconnected critical infrastructure assets. Distribute Electricity is at greatest risk from flooding, tropical cyclones and hurricanes, and extreme heat, which threaten the equipment at distribution substations, utility poles, and power conductors that serve as key nodes for distributing electricity throughout the power grid.[2] Furthermore, substations have components that were not designed for the climate risks of the next century (Munir, Mustafa, and Siagian, 2019). For example, extreme heat threatens distribution equipment that is not rated for safe operation above certain design thresholds. Relevant design standards for these assets were not created with consideration of future changes in extreme heat (Agrawal, 2020; Doherty and Dewey, 1925; Nateghi, 2018; U.S. Department of Energy, 2015; Wuebbles et al., 2017). Increased intensity and frequency of climate driver events may stress response operations and increase maintenance demands.

We identified 24 unique strategies, heavily focused on physical damage or disruption, that address the six impact pathways relevant to this NCF (Table 3.3). For example, implementing hardening measures, increasing reserve margins for generation capacity, maintaining backup systems, improving planning and monitoring nationally and at the system level, and maintaining wetlands and other natural areas as part of floodplain management can help reduce the potential for physical damage. Limiting the effect of extreme heat on demand for electricity distribution can be addressed through strategies that enhance demand-side efficiencies and reliabilities, either through implementing technical approaches; regulatory, education, or behavioral incentives; or demand-sensitive tariffs.

We assessed that five of the 24 strategies would increase cyber vulnerabilities. These were

- developing advanced visualization and information systems
- deploying distributed generation (distributed PV, microgrids, and minigrids)
- developing smart grids
- employing improved hazard mapping and monitoring technology
- increasing generation capacity reserve margins
- planning an established alternative or on-site power supply.

We also assessed that three of the strategies could potentially reduce cyber vulnerabilities (enhanced supply-side efficiency and reliability, improve integrated electricity planning approaches, and more-efficient resource use).

---

[2] The other electricity-related NCFs are Transmit Electricity and Generate Electricity.

TABLE 3.3

Adaptation Strategies Addressing Distribute Electricity Impact Pathways

Strategy	Demand Change	Physical Damage or Disruption
Build sea walls and coastal protection structures		▪
Conserve and restore wetlands and floodplains		▪ ▪ ▪
Deploy distributed generation: distributed PV, microgrids, minigrids		▪ ▪ ▪
Develop advanced visualization and information systems		▪ ▪ ▪ ▪
Develop and build smart grids		▪ ▪ ▪
Elevate electrical infrastructure		▪ ▪ ▪
Employ bushfire reduction and prescribed fire		▪
Employ hardening measures		▪ ▪ ▪ ▪
Enhance demand-side efficiency and reliability	▪	
Enhance resource efficiency	▪	
Enhance supply-side efficiency and reliability	▪	
Equipment design standards		▪
Implement demand-sensitive electricity tariffs	▪	
Improve climate value at risk assessment for utilities and assets		▪ ▪ ▪ ▪
Improve hazard mapping and monitoring technology	▪	▪ ▪ ▪ ▪
Improve integrated electricity planning approaches	▪	▪ ▪ ▪ ▪
Increase generation capacity reserve margins	▪	
Leverage national and regional adaptation plans	▪	▪ ▪ ▪ ▪
Maintain backup components and materials		▪ ▪ ▪ ▪
Perform integrated coastal zone management		▪
Perform vulnerability assessments		▪ ▪ ▪ ▪
Relocate electrical infrastructure		▪ ▪ ▪
Strengthen utility mutual aid agreements		▪ ▪ ▪
Clear transmission line rights-of-way		▪
Install underground distribution conductor		▪ ▪
Improve underground distribution conductor code		▪ ▪

▪ Extreme heat   ▪ Hurricanes   ▪ Wildfire

▪ Flooding   ▪ Sea-level rise

A full discussion of each NCF was not practicable within project resources; however, Appendix A presents the unique climate adaptation strategies, their sources, and their ratings for each NCF. The remainder of this chapter presents information that summarizes across NCFs for those in government or the private sector who may be interested in using the NCF framework to glean general observations about the distribution of climate adaptation strategies across the impact pathways. Specifically, the information can be used to understand where there are strategies that have general utility across a number of impact pathways, assess which NCFs have many options for mitigating climate risk for a given impact pathway, reveal where opportunities may be more limited, or reveal where there are fewer highly effective or feasible strategies. This information could be useful to policymakers and others involved with climate risk mitigation.

Characterization of Adaptation Strategies Identified Across All 25 High-Risk NCFs

This section includes observations about the full set of climate adaptation strategies for the 25 NCFs at risk of moderate disruption from climate change. We identified climate adaptation strategies that addressed each of the 179 impact pathways that led to a risk of a moderate disruption by 2100.

The matrix showing all the impact pathways is presented in Table 3.4. As discussed in Figure 3.1, each row in the table represents an NCF, while the columns refer to impact mechanisms and the color of the squares indicate the climate drivers. Our previous report (Miro et al., 2022) found that flooding, sea-level rise, and tropical cyclones and hurricanes pose the greatest risk of disruption to the assessed NCFs at the national level and recommended that strategies to address these risks be prioritized for investment.

Mapping and Distribution of Climate Adaptation Strategies

We mapped a total of 967 adaptation strategy–impact pathway combinations to mitigate risk to the NCFs (allowing strategies to be repeated across multiple pathways). These high-level strategies are commonly mentioned in the literature. While other strategies may be applicable to adapting to climate change, these emerged from the literature review the HSOAC team SMEs conducted during the limited time frame for this study.

Building off the previous table, Table 3.5 builds on Table 3.4 by adding the number of strategies identified for each impact pathway (shown within each of the squares).

Table 3.5 shows that the number of climate adaptation strategies found for each NCF climate driver–impact pathway varies. **Several NCFs emerged as potentially strategy-rich, with ten or more adaptation options identified for several impact pathways.** Twenty-eight impact pathways have more than ten options for mitigating risks due to climate change, and the most for a single pathway is 17 for physical damage or disruption from sea-level rise to the Transmit and Distribute Electricity NCFs. The Energy and Transportation sectors and Provide Housing emerged as potentially strategy-rich areas for mitigating physical damage and may be promising areas for investment given the number of potential options, although this is less true for input or resource constraints and for workforce shortages. Given that Distribute Electricity has the largest number of NCFs that are dependent on it, it may be particularly worthy of consideration for investment.

In general, SMEs reported that it was easier to identify strategies that mitigated risk of physical damage and harder to find those that worked against a specific input or resource constraint; however, this varied by NCF. For example, the Government and Social Services sector is strongly affected by changes in demand for the service after a climate driver event, suggesting that overall preparedness and risk mitigation activities by the public are important for reducing climate risk to this sector.

In contrast, for 43 impact pathways (approximately a quarter of the 179 impact pathways), we identified only one option for mitigating climate risk. Table 3.6 shows the subset of impact pathways for which

TABLE 3.4
Matrix of 179 Impact Pathways

Sector	NCF	Impact Mechanism			
		Physical Damage or Disruption	Demand Changes	Input or Resource Constraints	Workforce Shortages
Agriculture	Produce and Provide Agricultural Products and Services				
Energy	Distribute Electricity				
	Exploration and Extraction of Fuels				
	Generate Electricity				
	Transmit Electricity				
Government and Social Services	Educate and Train				
	Enforce Law				
	Prepare for and Manage Emergencies				
	Provide Housing				
	Provide Medical Care				
	Provide Public Safety				
Industry	Maintain Supply Chains				
	Manufacture Equipment				
	Produce Chemicals				
	Provide Insurance Services				
Infrastructure	Develop and Maintain Public Works and Services				
	Provide and Maintain Infrastructure				

Table 3.4—Continued

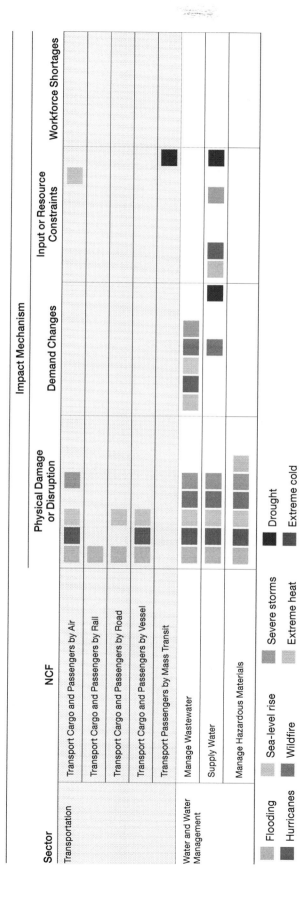

TABLE 3.5
Number of Adaptation Strategies, by NCF, Climate Driver, and Impact Mechanism

Sector	NCF	Physical Damage or Disruptions	Demand Changes	Input or Resource Constraints	Workforce Shortages
Agriculture	Produce and Provide Agricultural Products and Services	5 6 4 5 7 4 3		1 1 1 1 4 16	3
Energy	Distribute Electricity	14 15 17 3 12	10		
	Exploration and Extraction of Fuels	7		5	
	Generate Electricity	14 15 1 10 8	9 6 9	8 7	
	Transmit Electricity	14 15 17 3 12			
Government and Social Services	Educate and Train	5 3 5 6 3	2 2 2		2
	Enforce Law	1 1 1 1 1	1 2 1 2 3	1	3 3 3 3
	Prepare for and Manage Emergencies	1 1 1 1 1	3 3 3 3		
	Provide Housing	12 13 13 10 8 11	11 11 11 7 9		
	Provide Medical Care	7 7 7 6 7	3 3 3 3		
	Provide Public Safety	1 1 1 1 1	3 3 3 3		
Industry	Maintain Supply Chains	4 3 2 3 1	3	3 4 1 1	3
	Manufacture Equipment	4 3 2 3 1		3 4 1	1
	Produce Chemicals	4 3 2			
	Provide Insurance Services	2 2 6	2 2 2		
Infrastructure	Develop and Maintain Public Works and Services	8 4 7 4 4 7 3 2	4 3	2 5	3 1
	Provide and Maintain Infrastructure	8 7 9 3 4 7 2 2		2 1 1 2	2 1 1 1
Transportation	Transport Cargo and Passengers by Air	12 13 8 13		3	
	Transport Cargo and Passengers by Rail	16			
	Transport Cargo and Passengers by Road	15 12			
	Transport Cargo and Passengers by Vessel	16 15 13			
	Transport Passengers by Mass Transit			4	

Impact Mechanism

Table 3.5—Continued

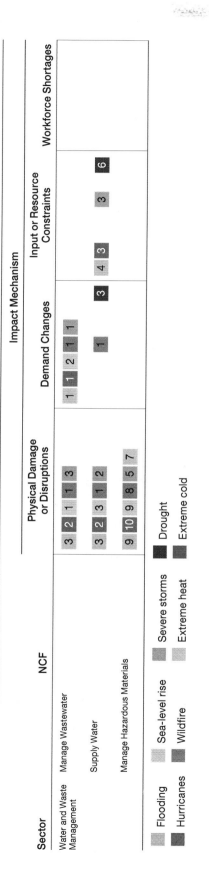

Sector	NCF	Impact Mechanism			
		Physical Damage or Disruptions	Demand Changes	Input or Resource Constraints	Workforce Shortages
Water and Waste Management	Manage Wastewater	3 2 1 1 3	1 2 1 1		
	Supply Water	3 2 3 1 2	1 3	4 3 3	
	Manage Hazardous Materials	9 10 9 8 5 7		6	

Flooding Sea-level rise Severe storms Drought

Hurricanes Wildfire Extreme heat Extreme cold

we were able to identify only a single strategy. Twelve NCFs in the Agriculture, Energy, Government and Social Services, Industry, Infrastructure, and Water and Waste Management sectors have only one climate risk mitigation option for several climate driver–impact mechanism combinations. As discussed in Chapter Two, SMEs were asked to identify at least one adaptation strategy per impact pathway; without this guidance, some impact pathways might not have had any adaptation strategies identified. While the quantity of strategies available for any single pathway does not necessarily indicate quality (although prevalence in the literature is likely positively associated with quality), having only one option is potentially worrisome if it is not feasible to implement in all contexts (e.g., specific location, community).[3] For example, as mentioned earlier, a high-priority NCF, Provide Public Safety, has four feasible strategies, all of which work to increase workforce capacity. However, financial constraints or limited availability of qualified personnel at the local level could preclude implementing them. While not universally true, intuitively, at the national level, having more options rather than fewer provides greater flexibility to reduce risk, given that local vulnerabilities, capabilities, and contexts can vary dramatically in ways that influence both the feasibility and effectiveness of a given strategy.

The climate driver with the most adaptation strategy–impact pathway combinations is flooding, with 230 combinations. As Table 3.5 shows, nearly all 25 NCFs were assessed as being at moderate-level risk from flooding (only Exploration and Extraction of Fuels and Transport Passengers by Mass Transit were not assessed at this risk level for flooding). This may be unsurprising, given the prevalence of flooding and flood risk in the United States. Floods from riverine and flash flooding produced by high rainfall events could cause major disruption to infrastructure systems through direct damage from floodwaters or debris, loss of access to assets, and cascading effects within infrastructure systems. Because flooding is so common over most of the United States, this climate change driver is expected to have effects across the country. Flooding can inflict significant physical damage and disruptions on NCFs with large physical footprints, which is likely why nearly 80 percent of the adaptation strategies for flooding targeted to owner-operators seek to limit physical damage or disruption.

However, because the extent and location of flooding are highly influenced by land-use patterns, additional strategies may be implemented to enhance or complement those that seek to prevent physical damage from water inundation. Development, farming, and other land uses affect the flow of water and the transport of sediment and pollutants. Furthermore, because land-use patterns can either accelerate or absorb water flow, actions in one community or geographic area will affect neighboring areas. Similarly, the location of critical assets within flood-prone areas influences the consequences of flooding. Comprehensive flood management therefore requires coordinated action by several actors in the community and region, in addition to owner-operators.

The climate driver with the second-highest number of strategy–impact pathway combinations is sea-level rise, with 169 combinations. Nearly all of the 25 NCFs are also at moderate risk from sea-level rise, as shown in Table 3.5. Because both the flooding and sea-level rise climate drivers can disrupt NCFs through temporary or permanent inundation of facilities, infrastructure assets, or essential infrastructure system components with water, the drivers share many of the same climate adaptation strategies (95) across multiple NCFs. Our previous report (Miro et al., 2022) found that flooding, sea-level rise, and tropical cyclones and hurricanes pose the greater risk of disruption to the assessed NCFs at the national level and recommended prioritizing strategies to address these risks for investment.

---

[3] We provide our assessment ratings for each strategy in Appendix A. However, as discussed throughout the report, these are high-level, general assessments, and the feasibility or effectiveness of an adaptation strategy in a specific community will be highly dependent on local conditions.

TABLE 3.6

Subset of Impact Pathways with Only One Identified Adaptation Strategy, by NCF, Climate Driver, and Impact Mechanism

Sector	NCF	Impact Mechanism			
		Physical Damage/ Disruptions	Demand Changes	Input/Resource Constraints	Workforce Shortages
Agriculture	Produce and provide agricultural products and services			1 1 1	
Energy	Generate Electricity	1			
Government and Social Services	Enforce Law	1 1 1	1	1	
	Prepare for and Manage Emergencies	1 1 1	1		
	Provide Public Safety	1 1 1			
Industry	Maintain Supply Chains	1			
	Manufacture Equipment	1		1	1
	Produce Chemicals	1		1	
Infrastructure	Develop and Maintain Public Works and Services				
	Provide and Maintain Infrastructure			1 1	
Water and Water Management	Manage Wastewater	1	1 1		1 1
	Supply Water	1	1		

Legend (climate driver):
- Flooding
- Hurricanes
- Sea-level rise
- Wildfire
- Severe storms
- Extreme heat
- Drought

The climate driver with fewest strategies across NCFs and mechanisms is extreme cold, with four combinations. Extreme cold is the only climate driver expected to decrease between now and 2100. As a result, extreme cold events are not expected to cause substantial disruption at the national scale for the NCFs studied, requiring fewer adaptation strategies. Even though recent extreme cold events have shown NCFs' sensitivity to extreme cold,[4] demonstrating the severe consequences that can result at a regional scale, future climate forecasts do not suggest a significant possibility of comparable events in the future that could impact NCFs at the national scale.

We now review how much choice or flexibility owner-operators may have to mitigate risks driven by climate change. Table 3.7 characterizes how the climate adaptation strategies we identified are distributed among the climate drivers and impact mechanisms. It shows the number of unique strategies that could potentially mitigate risk for each climate driver for each impact mechanism. If a strategy is relevant to multiple drivers, it is replicated in the respective rows for each driver, resulting in a total greater than 254. (For example, as discussed earlier, the flooding and sea-level rise drivers share many of the same strategies.) As Table 3.7 illustrates, we identified significantly more options for mitigating physical damage or disruption than for the other mechanisms. As a first-order screening, this suggests that there are a reasonable number of options for mitigating climate risk. However, as noted previously, quantity alone is not necessarily indicative of a superior set of options; strategies are not necessarily equivalent in terms of focus, quality, and scale, all of which influence effectiveness.

Adaptation Strategies That Address Risk from Multiple Impact Pathways

Adaptation strategies that were assessed to be relevant to multiple NCFs, climate drivers, or impact mechanisms have the potential to provide greater benefits for a given level of investment.

TABLE 3.7
Adaptation Strategy Count, by Climate Driver and Impact Mechanism

Climate Driver	Physical Damage/ Disruptions	Demand Changes	Input/Resource Constraints	Workforce Shortages
Flooding	123	19	18	4
Sea-level rise	94	9	1	—
Tropical cyclones and hurricanes	89	21	9	4
Extreme heat	47	24	12	9
Wildfire	46	18	1	4
Severe storm systems (nontropical)	48	14	4	3
Drought	18	12	33	—
Extreme cold	2	—	—	—

NOTE: Counts in each cell indicate the unique number of strategies for that climate driver, although the counts may replicate strategies across drivers.

[4] An extreme cold event in February 2021 left more than 4.5 million customers, or more than 10 million people, in Texas without electricity, in some cases for several days. See Busby et al., 2021.

Strategies That Address Multiple NCFs

We identified 28 strategies in the literature that apply to four or more NCFs across climate drivers and impact mechanisms. The adaptation strategy that applies to the most NCFs is sea walls and coastal protection structures, which applies to Generate Electricity, Distribute Electricity, Transmit Electricity, Maintain Supply Chains, Manage Hazardous Materials, Provide and Maintain Infrastructure, Manufacture Equipment, Produce Chemicals, and Provide Housing. A sea wall is a solid-built structure that is used to separate the land from the sea. Sea walls protect the bases of cliffs, land, and buildings against erosion and can prevent coastal flooding in many regions. However, sea walls can be expensive to build and maintain and, depending on their location, may disrupt natural processes, such as habitat migration or sediment flow, that can exacerbate beach erosion. They can also cause "coastal squeeze," leading to the loss of natural defensive barriers, such as wetlands, which can, in turn, increase the risk to the NCFs. The NCFs with coastal infrastructure are subject to sea-level rise, which Table 3.8 shows is one of only two climate drivers to which this strategy applies. The strategy applicable to the second-highest number of NCFs is vulnerability assessments, which is applicable to Generate Electricity, Distribute Electricity, Transmit Electricity, Manage Hazardous Materials, Exploration and Extraction of Fuels, and Provide Housing.

Strategies That Address Multiple Climate Drivers

We identified 14 strategies that address at least six of the eight climate drivers (Table 3.9). We assessed these 14 unique climate adaptation strategies as applying to five NCFs: Develop and Maintain Public Works and Services, Manage Hazardous Materials, Produce and Provide Agricultural Products and Services, Provide and Maintain Infrastructure, and Provide Housing. Many of these strategies are related to assessing, planning, and preparedness, such as "scenario planning, vulnerability assessments, and community-based adaptation actions," "improved eco-system management and land-use approaches," and strategies to "strengthen building codes and standards to reflect changing climate conditions." Many of these strategies can be and are being pursued today because they involve integrating climate risk assessment and mitigation into existing activities. A majority of these address physical damage/disruption mechanisms, which is not surprising because the NCFs were largely vulnerable to this climate change mechanism.

Strategies That Address Multiple Impact Mechanisms

We identified 40 unique strategies as helping to mitigate climate risk across more than one impact mechanism, with nine strategies that were applicable to three mechanisms (Table 3.10). Although we did not consider strategy implementation scope or scale, these climate adaptation strategies may have the potential to either be more effective for a given level of investment or provide synergistic benefits with other strategies because they address multiple impact mechanisms. For example, several strategies relevant to the Housing NCF work to reduce both physical damage from extreme heat and changes to demand for housing. Such strategies as building insulation, passive cooling, mechanical ventilation systems, social safety nets and protections, green infrastructure, and natural areas can reduce vulnerabilities to extreme heat for physical damage or disruption and demand changes. Table 3.10 also suggests that addressing risk from drought may require a portfolio of activities to manage climate change–related risk. Because these strategies were gathered by NCF and the primary climate driver and impact mechanism, additional analysis could reveal additional linkages between a strategy and other climate driver and mechanism combinations. For example, improved hazard mapping and monitoring should likely apply to all climate drivers, as should workforce education, which would likely help address any climate driver event that produces a workforce shortage.

TABLE 3.8

Adaptation Strategies Addressing Four or More NCFs, by Impact Mechanism and Climate Driver

Strategy	Physical Damage or Disruptions	Input or Resource Constraints	Demand Changes	Workforce Shortages
Build a volunteer reserve corps			■■■■■	
Build levees and dikes	■	■		
Build sea walls and coastal protection structures	■■	■		
Conserve and restore wetlands and floodplains	■■■			
Deploy distributed generation: distributed PV, microgrids, minigrids	■■■	■		
Develop advanced visualization and information systems	■■■■■			
Develop and build smart grids	■■■	■		
Develop and exercise mutual aid agreements to cope with demand surge during disasters			■■■■■	
Develop workplace heat standards	■	■		■
Elevate electrical infrastructure	■■■	■		
Employ blockchain technology to increase supply chain transparency	■■	■■		
Employ bushfire reduction and prescribed fire	■			
Employ hardening measures	■■■■			
Enhance demand-side efficiency and reliability			■	
Enhance resource efficiency			■	
Enhance supply-side efficiency and reliability			■	
Enhance zoning and land use, and relocate critical assets in vulnerable areas	■■■			
Implement demand-sensitive electricity tariffs			■	
Improve integrated electricity planning approaches	■■■■■		■■	
Improve climate value at risk assessment for utilities and assets	■■■■■	■■		
Improve hazard mapping and monitoring technology	■■■■■	■■	■■	
Leverage national and regional adaptation plans	■■■■■	■■	■■	
Maintain backup components and materials	■■■■			
Perform integrated coastal zone management	■			
Perform vulnerability assessments	■■■■■			

Table 3.8—Continued

Strategy	Physical Damage or Disruptions	Input or Resource Constraints	Demand Changes	Workforce Shortages
Relocate electrical infrastructure	▦ ▦ ▦ ▦	▦		
Stockpile critical goods		▦	▦	
Strengthen utility mutual aid agreements	■ ▦ ▦ ▦ ▦	■ ▦		

■ Drought   ▦ Flooding   ▦ Sea-level rise   ▦ Wildfire
▦ Extreme heat   ▦ Hurricanes   ▦ Severe storms

TABLE 3.9
Adaptation Strategies Addressing Six or More of the Eight Climate Drivers, by Mechanism

Strategy	Impact Mechanism			
	Physical Damage or Disruption	Demand Changes	Input or Resource Constraints	Workforce Shortages
Adapt infrastructure to manage and store land resources	■ ▦ ▦ ▦ ▦ ▦		▦	
Adopt financial mechanisms (e.g., mutual aid agreements or insurance policies)	■ ▦ ▦ ▦ ▦ ▦ ▦			
Carry out scenario planning, vulnerability assessments, and community-based adaptation actions	▦ ▦ ■ ▦ ▦ ▦	▦ ▦ ■ ▦ ▦		
Enhance insurance and related products	▦ ▦ ■ ▦ ▦			
Enhance social safety nets and protections	▦ ▦ ■ ▦ ▦	▦ ▦ ■ ▦ ▦		
Enhance zoning and inspections	▦ ▦ ■ ▦ ▦			
Implement ecological restoration, preservation, conservation, and natural resource management	■ ▦ ▦ ■ ▦ ▦ ▦		■ ▦	
Implement technologies to monitor and mitigate climate impacts	▦ ▦ ■ ▦ ▦ ▦		■ ▦	
Improve government planning, preparedness, and/or investment	▦ ▦ ■		■ ▦ ■ ▦	▦
Incorporate redundancies and/or decentralized options (e.g., microgrids, battery storage)	■ ▦ ▦ ■ ▦ ▦			
Invest and implement disaster preparedness and response	▦ ▦ ■ ▦ ▦	▦ ▦ ■ ▦ ▦		
Strengthen building codes and standards	■ ■ ▦ ▦ ■ ▦ ■			
Support household preparation	▦ ▦ ■ ▦ ▦			
Use alternative materials to withstand and/or adapt to future changing climate conditions	■ ▦ ▦ ■ ▦ ▦			

■ Drought   ▦ Extreme heat   ▦ Hurricanes   ▦ Severe storms
■ Extreme cold   ▦ Flooding   ▦ Sea-level rise   ▦ Wildfire

TABLE 3.10

Adaptation Strategies Addressing Three Impact Mechanisms

Strategy	Impact Mechanism			
	Physical Damage/ Disruptions	Input/Resource Constraints	Demand Changes	Workforce Shortages
Develop workplace heat standards	■	■		■
Educate agricultural workforce and raise awareness for agricultural producers	■ ■ ■	■		■
Improve government planning, preparedness, and/or investment	■ ■ ■	■ ■ ■ ■		■
Improve hazard mapping and monitoring technology	■ ■ ■ ■ ■	■ ■	■ ■	
Leverage national and regional adaptation plans	■ ■ ■ ■ ■	■ ■	■ ■	
Manage supply chains	■	■	■	
Regulate key resources and enact laws to support risk reduction	■ ■	■ ■		■
Substitute renewable energy technology	■	■	■	
Utilize and enhance municipal water management programs	■	■	■	

■ Drought	■ Sea-level rise	■ Wildfire
■ Flooding	■ Severe storms	
■ Extreme heat	■ Hurricanes	

Effectiveness and Feasibility of Adaptation Strategies

We next discuss the results of our assessment of the likely effectiveness and feasibility of each strategy, how well the strategy was supported by evidence, and whether it might increase or reduce cyberattack vulnerabilities. The value of a strategy does not lie just in its relevance but also depends on its effectiveness and feasibility and on the strength of the evidence supporting a particular strategy. Figure 3.2 summarizes our assessment across the four categories described above for the 254 unique strategies we identified.

In general, the strategies we identified have medium to strong evidence to support them. In some cases, analysts selected strategies for which evidence was not as strong to ensure that at least one strategy was identified for each NCF–climate driver–mechanism combination. Only 5 percent of identified strategies had weak strength of evidence.

We assessed the majority of strategies as having moderate impact. This suggests that multiple strategies or packages of strategies may be needed if the goal is to minimize the effects of climate change on an NCF, particularly for those at the greatest risk. As discussed in Chapter Two, adaptation strategies rated as having no impact for a particular impact pathway were screened out and are not included in the final dataset.

Strategy feasibility ratings were mixed. We assessed approximately 47 percent of the climate adaptation strategies to have high feasibility, with the remaining 53 percent of medium or low feasibility. Lower ratings can often be attributed to the anticipated scale or resources required to successfully implement the strategy. For example, "national and regional adaptation plans" are cited as a climate adaptation strategy with strong strength of evidence and major impact but low feasibility. This is likely because the United States has never had a national adaptation plan, and establishing national coverage of the NCFs with regional adaptation plans would be a complex process requiring the participation of many different actors. While low feasibility does not imply that the climate adaptation strategy is impossible, the rating does mean that the SME found the barriers to success to be significant.

FIGURE 3.2

Summary of Adaptation Strategies' Strength of Evidence, Effectiveness, Feasibility, and Cyber Vulnerability

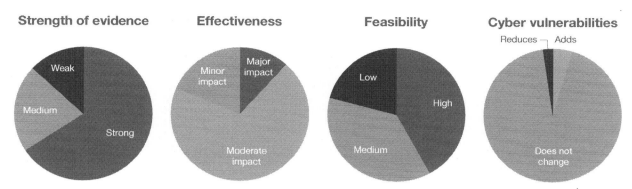

We assessed that only a small portion of the strategies would impact cyber risk either positively or negatively. Those that were assessed to add cyber vulnerabilities include many that require networked information systems to monitor environmental or operating conditions or to control distributed systems:

- develop advanced visualization and information systems
- cyber investigations and/or monitoring of federal and other assistance programs and funds
- deploy distributed generation: distributed PV cells, microgrids, and minigrids
- develop smart grids
- implement early warning systems
- improve government planning, preparedness, and/or investment
- implement technologies to monitor and mitigate climate impacts (for agricultural systems)
- improve infrastructure monitoring (including in real time)
- improve river condition monitoring
- improve vehicle communications technology
- improve weather detection and communication systems
- employ improved hazard mapping and monitoring technology
- increase generation capacity reserve margins
- plan and establish alternative or on-site power supply
- provide real-time information to drivers
- substitute renewable energy technology substitution
- strengthen online and remote instruction capabilities.

The strategies that we assessed to reduce cyber vulnerabilities include several related to planning and vulnerability assessments, adding system redundancies, developing improved codes and standards, employing zoning and inspections, and mapping supply chains.

The tables in Appendix A show the distribution of adaptation strategies by sector and NCF and SME ratings for each assessment category (strength of evidence, effectiveness, feasibility, and cyber vulnerability). The strategies summarized in this report and included in Appendix A are high-level strategies that are commonly mentioned in the literature for the impact pathways at greatest risk of climate change effects. While other strategies may be applicable to adapting to climate change, these were identified during the literature review conducted by the HSOAC team of analysts with subject-matter expertise during the limited time frame scoped for this study.

CHAPTER FOUR

Considerations in Selecting an Adaptation Strategy

To this point, the discussion in the report has centered on providing information on climate adaptation strategies across NCFs to describe strategies that have general utility across several impact pathways, assess which NCFs have many options for mitigating climate risk for a given impact pathway, and reveal where opportunities may be more limited or where there are fewer highly effective or feasible strategies. This information, combined with the list of strategies in Appendix A and the forthcoming tool, could be useful to owner-operators, policymakers, and others as a starting point for identifying options to mitigate the risk to NCFs from climate change.

We conducted a general, high-level assessment of each strategy's effectiveness, feasibility, strength of evidence, and impact on cybersecurity for a given impact pathway. These assessments may provide an initial basis for identifying which adaptation strategies could potentially be advantageous, although investment decisions for a specific entity or community will need to be based on more-precise assessments that account for the specific context in which the strategy would be implemented, as well as a host of other considerations.

Ideally, climate adaptation strategies should be considered within established planning processes, such as capital improvement or land-use plans, and should involve stakeholders in the development and assessment process. Adaptation strategies are more effective when part of a portfolio of actions. Investment decisions should consider co-benefits, assess both short- and longer-term risks, and recognize that the severity of these natural hazards is increasing in most places but that the precise changes and effects are uncertain.

In this chapter, we provide an overview—but by no means an exhaustive list—of several additional factors decisionmakers may wish to consider when selecting among alternative adaptation strategies. We focus on local priorities and resources, interdependencies and co-benefits, uncertainties, and equity issues. We also discuss some tools that are available to help in selecting strategies.

Local Priorities and Resources

Selecting an adaptation strategy will depend on the priorities, needs, preferences, and resources of the entity implementing the strategy. The entity implementing the strategy may also be a community or a diverse collection of entities. Some strategies, while potentially effective, may not be consistent with other community goals or may exceed available resources. For example, many of the strategies we identified involve changing the built environment. Communities may have historical, economic, or other reasons for being hesitant to implement significant physical changes, such as the construction of a sea wall, in certain areas (for example, beaches that provide a source of income or recreation to the community). In other cases, while a strategy may be known to be effective, it may simply be prohibitively expensive for an entity to implement. One example of this might include moving equipment underground. While a strategy may be proven to be effective at reducing risk to a particular NCF, it may simply be too costly for a community to implement on its own.

Interdependencies and Co-Benefits

Adaptation strategy selection should also consider interdependencies between the strategies and between NCFs. As discussed in Chapter Three, several of the strategies we identified may address risk for more than one impact pathway or may address risk to multiple NCFs. Holding other factors, such as cost and feasibility, equal, these strategies may be preferable to strategies that do not produce co-benefits. Investments in climate adaptation strategies may also provide additional benefits that should be considered in the decisionmaking process. For example, natural infrastructure mitigates some climate risk while providing the co-benefits of ecosystem services to the community.[1] However, benefits of this type may be challenging to distill into a single value for a traditional cost-benefit analysis. Decisionmaking processes should reflect this complexity and be based on effectiveness in reducing climate risk and on increasing any potential co-benefits, such as those to the economy, environment, or social systems.[2]

Adaptation strategies are also often more effective when they are part of a portfolio of actions. Investments in climate adaptation strategies may be made by individuals; local, state, regional, or national government organizations; or private-sector entities of any size. Strategies can be implemented at different scales and levels of investment. Ideally, adaptation strategy selection would be considered within established planning processes that balance many societal objectives, such as capital improvement or land-use plans, and would involve multiple stakeholders to ensure decisions incorporate these objectives and stakeholder values.

Finally, many NCFs are interdependent with others. As discussed in Chapter Three, in the risk assessment phase of this project, we found that the Distribute Electricity NCF is a critical upstream NCF to 20 of the 25 highest-risk NCFs.[3] As a result, disruption or failure of Distribute Electricity could have significant cascading and immediate effects on multiple NCFs. This may suggest that strategies that address risk to Distribute Electricity should be given greater weight than those that do not. More fundamentally, in addition to prioritizing adaptation strategies, it may be useful for businesses, governments, and communities to prioritize certain NCFs or subfunctions based not only on their direct needs from individual NCFs but also on interdependencies among NCFs.

Uncertainties

While there is strong evidence that climate change is occurring, there is uncertainty about the magnitude, frequency, timing, and location of the effects of natural hazards. The results of implemented adaptation strategies may also be uncertain and depend on various factors, including the scale and quality of implementation. Given these uncertainties, decisions about adaptation strategy selection should, ideally, use a form of iterative risk management, which uses an assessment, monitoring, learning, and adjustment cycle to manage risk. An example of an iterative cycle would be to assess vulnerabilities from climate change under a range of plausible futures, make investment decisions on this basis, continually monitor risk to incorporate changes in understanding of climate change and real-world manifestation of risk, and adjust investments accordingly.

---

[1] *Ecosystem services* are the benefits that natural areas provide. For example, forests help improve air quality and reduce sediment into streams, and wetlands can absorb runoff or storm surges and protect communities or infrastructure from flooding.

[2] An in-depth discussion of decisionmaking in uncertainty is beyond the scope of this research, but interested readers may review Groves et al., 2019; Lempert et al., 2020; Marchau et al., 2019; U.S. Climate Resilience Toolkit, undated; or Adaptation Clearinghouse, 2022.

[3] Please see Miro et al., 2022, for a more complete discussion of NCF interdependency. An *upstream NCF* is defined as one that provides a key input to another NCF.

Some strategies may be classified as "no regrets," which refers to the likelihood that the investment would be beneficial regardless of the future magnitude of climate change effects. (An example of this might be avoiding construction in existing flood-prone areas.) When there is substantial uncertainty, as there is in dealing with climate change–related risk, investment decisions that employ strategies that are flexible or adaptable to changing conditions are generally preferable. These types of strategies may include using less-expensive, shorter-lived strategies (e.g., temporary shelters) or adding larger design safety margins that have minimal cost (e.g., larger culverts). Other approaches to managing uncertainty that may be relevant to climate change–related risk include developing thresholds based on some variable related to the climate that trigger investment in a specific adaptation strategy (U.S. Department of the Navy, 2017; Hallegatte et al., 2012; Groves et al., 2019; Lempert et al., 2020; Lempert, 2014; Lempert and Groves, 2010; Marchau et al., 2019).

Equity and Unintended Consequences

Adaptation strategy investment decisions should also consider the potential for inequitable impacts on vulnerable populations. Assistance to stakeholders considering different adaptation strategies should include support to identify and consider wider social and economic factors that influence vulnerability (Levine, 2004) and resilience, which are outside the current risk assessment framework. *Vulnerability* has been defined as the "conditions determined by physical, social, economic and environmental factors or processes, which increase the susceptibility of a community to the impact of hazards" (International Strategy for Disaster Reduction, 2005, p. 1). These conditions can range from infrastructure-focused conditions, such as deterioration of power and water systems, to human-focused conditions, such as weak social support networks. The Hyogo Framework 2005–2015 (International Strategy for Disaster Reduction, 2005) establishes the importance of an integrative approach to vulnerability from climate change that recognizes that climate variability and extreme events contribute to vulnerability of impoverished populations because they are more likely to depend on natural resource–based livelihoods and live in areas at high risk for natural disasters. Risk reduction decisionmaking that relies only narrowly on NCF-specific risk assessment may be shortsighted, may not account for community vulnerabilities that are hazard independent, and may result in unintended consequences if not considered during mitigation and adaptation planning.

Focusing only on the physical, input, demand, and workforce vulnerabilities of NCFs related to specific climate change events may overlook drivers of community resilience and sustainable development and may make it more likely for adaptation strategies to have unintended consequences.[4] These unintended consequences are often termed *maladaptation* and occur when "intervention in one location or sector could increase the vulnerability of another location or sector or increase the vulnerability of the target group to future climate change (medium evidence, high agreement)" (Noble et al., 2014, p. 837), as well as current climate change impacts. Exacerbating or redistributing existing vulnerabilities or creating new sources of vulnerability is especially problematic for low-income groups and communities of color, which often bear a heavier burden of health risks and impacts from climate change (e.g., are more likely to live in polluted areas or areas with fewer available services). These unintended consequences result not only from poorly planned adaptation actions but also from risk-reduction decisionmaking that emphasizes short-term outcomes of an

---

[4] While *community resilience* has been defined in myriad ways, most definitions notably settle on some combination of a community's ability to withstand, recover, and adapt to adversity (Chandra et al., 2011). *Sustainable development* is most widely defined as ensuring that development "meets the needs of the present without compromising the ability of future generations to meet their own needs" (World Commission on Environment and Development, 1987). However, while definitions vary widely, there is agreement that sustainable development sits at the intersection of three overlapping concepts: environment, development, and equity. See Robert, Parris, and Leiserowitz, 2005.

adaptation strategy and discounts the full range of interactions that can result from implementation (Noble et al., 2014). Thus, advances in climate adaptation planning have begun to focus on factors beyond biophysical vulnerabilities and include underlying causes of community vulnerability and resilience.

Resilience refers to a system's inherent ability to recover its functionality after events and has been applied to a community's recovery ability after disasters. Community resilience explicitly acknowledges hazard-independent vulnerabilities driven by socioeconomic and demographic factors (e.g., unequal access to a clean environment and basic environmental resources based on race) (Acosta et al., 2017). While these vulnerabilities are described as being independent of hazards, the root causes and upstream drivers of these hazard-independent vulnerabilities are often the same as those that drive climate change. The systems that create the physical environment and associated community living conditions (e.g., energy, transportation, agriculture) are also the key contributors to climate change (e.g., Superfund sites are disproportionately located in disenfranchised communities). For example, consider increased ozone levels. One of the root causes of increased ozone levels is extreme heat. Extreme heat also increases community vulnerabilities (e.g., rates of chronic disease) for populations that live in urban heat islands. Latinos are 21 percent more likely than non-Latino whites to live in urban heat islands, dominated by heat-retaining asphalt and concrete and nearly half of all Latinos live in counties frequently violating clean air and ozone standards (Dodgen et al., 2016). Not surprisingly, this poor air quality contributes to disparate incidences of chronic disease. Specifically, Latino children are twice as likely to die from asthma as non-Latino whites. Extreme heat is a root cause contributing to both increased ozone levels (climate change) and increased mortality from asthma (community vulnerability) (Gamble et al., 2016).

As CISA assists stakeholders with evaluating and prioritizing climate adaptation strategies, there are a couple of considerations that can help reduce the likelihood that these strategies will have unintended consequences and increase the likelihood that these strategies will lead to more-resilient and -equitable communities. CISA should consider how to assist stakeholders to do the following:

- Develop a more complete understanding of vulnerability (hazard specific and hazard independent) and social inequities that contribute to community climate change risks to help improve understanding of the context for climate adaptation and provide insights into how strategies may increase or reduce existing vulnerabilities and social inequities. For future risk assessments, incorporate indicators of equity (e.g., percentage of people of color in a community, housing burden by tenure, severity, and race/ethnicity) and resilience (e.g., availability of social support) to help provide a broader contextual perspective on vulnerability.[5]
- Conduct an intentional analysis on the long-term effects and potential positive and negative spillovers on other areas and groups (Truelove et al., 2014). This goes beyond the cascade analysis done to date, to incorporate a more nuanced assessment of the interactional and dynamic impacts of climate change drivers and climate adaptation strategies on infrastructure *and* people—in particular, marginalized or underrepresented groups.
- Institute a process to lift up the perspectives of marginalized or underrepresented groups through authentic engagement to understand their priorities for design and implementation of adaptation strategies. If the trade-offs for and voices of these populations are not explicitly considered, making deci-

---

[5] An *equitable community* is defined as one where all people, regardless of race, gender, etc., are able to "fully participate in the community's economic vitality, contribute to its readiness for the future, and connect to its assets and resources" (PolicyLink and the USC Equity Research Institute, undated). An expanded definition of *equity* and example indicators can be found on the National Equity Atlas website (PolicyLink and the USC Equity Research Institute, undated). Example indicators of resilience are presented in National Research Council, 2015.

sions intended to be for the "greater good" could have the potential to result in the most significant and detrimental effects on the most vulnerable.

- Divorce climate adaptation from dominant infrastructure development agendas, which can overshadow how the specific climate drivers and societal factors drive vulnerability and concentrate land, capital, and resources in the hands of a few (Bulkeley and Tuts, 2013). Decisions should instead prioritize effectiveness of specific adaptation strategies to reduce vulnerabilities among the people and places that need it most and be accompanied by monitoring and evaluation to ensure that midcourse corrections can be made before strategies create maladaptation.

- Prioritize strategies that promote equitable adaptation to climate change.[6] *Many climate adaptation strategies have co-benefits that improve equity.* These win-win strategies include active transportation;[7] urban greening (parks and natural areas); mixed-use zoning; consumption of locally grown produce; and affordable, healthy, and energy-efficient housing.

- Measure the success of adaptation among the most vulnerable and include a focus on social vulnerabilities (Dilling et al., 2015). Averages and aggregates can mask effects on populations that have the most to gain from equitable climate adaptation strategies (Hsu et al., 2021).

While these six areas of consideration can begin to contribute to the prioritization of more equitable and resilient climate adaptation, CISA will ultimately need to assist stakeholders in conducting a more comprehensive and integrated approach to risk-reduction decisionmaking that moves beyond NCF-specific risk assessment. Prioritizing equitable climate adaptation strategies may not only help CISA meet its goals of protecting the NCFs but can also reduce existing inequalities.

Tools to Assist in Selecting Climate Adaptation Strategies

There are a vast number of decisions tools of varying quality and currency that could be used to guide vulnerability assessments and investment decisions. We began collecting those that had the most relevance to the NCFs and identified 50 guides and tools that owner-operators could use to assess alternative adaptation strategies. These are crosswalked against the NCFs for which they could be useful. As an example, Table 4.1 shows the decision-support tools identified for Distribute Electricity. The full list is provided in Appendix B.

Final Thoughts on Selecting Adaptation Strategies

In sum, selecting climate adaptation strategies must include considering potential short- and long-term effects, recognizing that the severity of natural hazards is increasing in most places but that the precise effects of climate change are uncertain. Therefore, investments should be flexible enough to meet changing conditions.

A best practice for managing climate risk is iterative risk management, which is the near-continuous process of assessing climate risk, investing in actions, monitoring and evaluating the results, and reevaluating opportunities for new actions that seeks to reduce climate risk either through reducing exposure to climate drivers, reducing the sensitivity to these drivers, or increasing adaptive capacity to limit damage (Lempert et al., 2018; Pörtner et al., 2022). Given the uncertainties around climate change effects both globally and

---

[6] Equitable adaptations are those that counter climate injustice—or the imbalance that those that have contributed the least emissions and have the least adaptive capacity are often disproportionately affected by climate change (Shi, 2021).

[7] For example, walking or biking that may be combined with public transportation.

TABLE 4.1

Sample of Identified Decision-Support Tools for Distribute Electricity and the Organizations They Might Benefit

Decision-Support Tools	Organization
Building America Solution Center—Disaster Resistance Tool	U.S. Department of Energy
Coastal Resilience Evaluation and Siting Tool (CREST)	National Fish and Wildlife Foundation
Community Resilience Portal	U.S. Department of Housing and Urban Development (HUD)
Community-Based Climate Adaptation Toolkit	Reef Resilience Network
EPA Flood Resilience Checklist	EPA
Managed Retreat Toolkit	Georgetown Climate Center
Natural Infrastructure Opportunities Tool	U.S. Army Corps of Engineers
Rapid Vulnerability & Adaptation Tool for Climate-Informed Community Planning	EcoAdapt
Regional Adaptation Collaborative Toolkit	Alliance of Regional Collaboratives for Climate Adaptation (California based)
Rolling Easements Primer	EPA
Synthesis of Adaptation Options for Coastal Areas	EPA
Urban Adaptation Assessment	Notre Dame Global Adaptation Initiative
U.S. Global Change Research Program (USGCRP) Federal Adaptation Resources website	USGCRP
Vulnerability, Consequences, and Adaptation Planning Scenarios	Social and Environmental Research Institute, Inc.

locally, decisionmakers should exercise caution when considering incremental solutions and should ensure that this does not preclude implementing more-transformative or -effective investments in the longer term as more information becomes available.

Finally, adaptation strategies should ideally be developed as part of a portfolio of actions to address climate risk and should involve all stakeholders in the development and assessment process. Investment decisions should avoid transferring risk to others in space and time and consider such co-benefits as improved economic opportunities, equities, or ecosystem services that are not easily distilled into a single numeric value. In Chapter Five, we summarize conclusions, limitations in our approach, and next steps for consideration.

Conclusions and Next Steps

This analysis gathers climate adaptation strategies across a broad set of functions that are critical to the economic prosperity, security, and well-being of the United States. The analysis is a step toward developing a more holistic understanding of the suite of adaptation strategies that can be used to meet the risks resulting from climate change, establishing potential priorities for these strategies, and understanding the interdependencies that exist across sectors or functions.

Key Takeaways

The NCFs provide a useful and still relatively new lens for assessing risk to the nation. As such, they are likely to reveal risks—and potential adaptation strategies—that may not be apparent through more-traditional frameworks for assessing critical infrastructure. One relevant example is the Provide Public Safety NCF. Traditional approaches to critical infrastructure have often focused on physical assets, such as buildings and other facilities. Under this type of framework, risk to fire, police, emergency medical services, and other public safety providers from climate change would primarily be assessed as risk to their physical facilities (for example, whether a fire station is located in an area that will become increasingly flood prone). By taking a functional approach to critical infrastructure, the NCFs expose additional risk to public safety agencies through increased demand on their workforces as a result of climate change—risk that we assess to be significant enough to disrupt the function on a regional basis as early as 2030.

Furthermore, the risk management framework (shown in Figure 1.1) developed for the risk assessment in Miro et al., 2022, provides a comprehensive and structured approach for understanding the sources of climate risk and their impact pathways to the NCFs. It can also be applied to expose second-order risks created by interdependencies and cascading effects with other NCFs, which can be useful when developing investment priorities. NCF owner-operators can use this approach to break down an unquestionably complex issue to better identify and target climate adaptation strategies for reducing the risk of climate change to their function(s).

We found that several adaptation strategies are applicable to most of the higher-vulnerability NCFs. However, in some cases, notably including Provide Public Safety—one of the two NCFs we assess as being at risk of a moderate disruption in 2030 (the other NCF is Supply Water)—we identified relatively few strategies. Furthermore, the adaptation strategies we identified for Provide Public Safety are concentrated on addressing the workforce to meet changing demand. This suggests that the ability of public safety agencies to implement these strategies remains in question. These and other priority NCFs remain areas in need of further investigation, novel solutions, and—more importantly—action.

The information in this report and the data collected provide a starting point for investment decisions. We identified 254 distinct adaptation strategies and at least one strategy for each impact pathway that we assessed to be vulnerable to moderate disruption as a result of climate change by 2100 (and, in many cases, much earlier). Applying the 254 unique adaptation strategies across the relevant impact pathways for the

NCFs at highest risk resulted in 967 adaptation strategy–impact pathway pairings, or potential solution spaces, to address these risks. The strategies can be sorted to address specific NCFs, climate drivers, or mechanisms of concern. They can be sorted by strength of evidence, feasibility, effectiveness, and cybersecurity concern. Collectively, they provide a starting point for risk mitigation planning that is aligned with the NCFs. In addition, the process can and should be repeated to further refine the set of climate adaptation strategies.

We assessed that, as a group, the adaptation strategies would generally not create additional cyber vulnerabilities, although this could change over time as new technologies emerge and as critical infrastructure systems are increasingly networked. However, CISA seems well positioned to assist owner-operators in addressing these vulnerabilities. We also assessed the effectiveness of each adaptation strategy, its evidence base, and how feasible it would be to implement, considering such factors as cost and the number of partners needed to implement the strategy. We identified examples of each strategy and citations for further information and currently available decision-support tools and provided our qualitative ratings for each strategy.

Our analysis focused on identifying strategies for mitigating the risks driven by climate change that were identified in our previous report (Miro et al., 2022). The identification of these strategies is just the beginning phase of the decisionmaking process. There are other considerations and trade-offs that decisionmakers and stakeholders must consider when prioritizing, selecting, and implementing adaptation strategies. These include, for example, interdependencies between NCFs and strategies and potential co-benefits or unintended consequences and negative impacts related to equity of the strategies.

Much remains to be done to support the range of actors and decisionmakers at the local level. Additional tools and resources are needed for incorporating adaptation options into decisionmaking, establishing priorities, and considering and incorporating stakeholder needs. These could include incorporating indicators of community vulnerability to improve decisionmaking and reduce the potential for maladaptation. In addition, while some tools exist, stakeholders are likely to benefit from additional locally relevant information and tools to identify, gather, and assess the information required to make facility-level decisions; assess costs, barriers, and enablers; and conduct cost-benefit and other analyses.

Next Steps

Our findings suggest several potential next steps.

Improve the candidate set of strategies. The structured approach we used can be repeated to identify new and emerging strategies and to improve assessments of feasibility, effectiveness, and strength of evidence. Specifically, future efforts could do the following:

- *Focus on identifying strategies for NCFs or NCF-impact pathways for which we did not identify a large number of strategies.* We identified only a single adaptation strategy for almost one-quarter of the 179 impact pathways for NCFs. These 12 NCFs and 43 impact pathways (shown in Table 3.6) may be candidates for future research efforts, particularly to the extent they reflect priority NCFs or climate drivers expected to significantly worsen in the future, such as Provide Public Safety and Supply Water. The NCFs that are the most vulnerable or have significant cascading interdependencies are clear priorities to ensure that viable options at the local level exist.
- *Explore strategies in areas we identified as potentially strategy rich.* Areas with a large number of strategies may be particularly ripe for investment. The Energy and Transportation sectors and the Provide Housing NCF each had large numbers of potential strategies, and one of the Energy sector NCFs, Distribute Electricity, is the NCF that has the largest number of NCFs that are dependent on it. Strategies in these areas should be explored and refined, including investigating how individual strategies could

be used in combination to produce greater risk reduction and identifying enablers and barriers to broad implementation.

- *Investigate strategies with broad applicability across NCFs.* Strategies that apply to and mitigate risk to multiple NCFs may be more likely to be cost-effective and to have broad-based support. These "bang for the buck" strategies should be prioritized for future research, outreach, technical materials, and assistance and, potentially, cross-NCF information-sharing.

- *Investigate why existing strategies have not been implemented at the scale required to reduce risk.* We identified several strategies with broad application across NCFs that have been in use for many years but assess that significant risk remains through the impact pathways to which these strategies apply. In many cases, these strategies have strong bases of evidence and high effectiveness ratings. Research should focus on what would be required to increase the results from these strategies otherwise proven to be effective, including implementation enablers and barriers.

- *Include additional generalized strategies and specific strategies for all stakeholders.* We did not include strategies that did not mention an NCF or climate change or were not relevant to the target audience of this report. However, these strategies may help address impact pathways for which we did not find multiple strategies. For example, household preparedness is a generalized strategy at the individual or family level but could significantly reduce strain on Provide Public Safety, one of the most at-risk NCFs. Strategies at this level may be particularly important for NCFs that provide public services or are heavily household-driven. Additionally, there may be other generalized strategies that reduce specific community vulnerabilities, which ultimately may reduce the risk to an NCF.

- *Regularly review and update the strategies to reflect new technology and emerging solutions.* New strategies and technologies focused on reducing the impacts of climate change are likely to appear on an increasing basis as climate risk manifests across the world. Scans should be conducted regularly to find and add these strategies. While we focused on more-proven strategies with evidence in the research literature that we assessed as at least moderately likely to be effective, future efforts could include more strategies with less of an evidence basis but with high potential payoffs with appropriate caveats.

Conduct future analyses that address this approach's limitations. Future analyses could address the limitations of this work. This could include accessing a broader array of sources, including proprietary ones; reviewing more-general adaptation strategies; including additional rating dimensions; and conducting more granular analyses of these and other strategy characteristics. Additional literature sources could include region-specific and professional literature. Assessment measures, such as feasibility and effectiveness, could be built out in a variety of ways, including addressing what factors influence these characteristics and how they might vary at the local level. Nevertheless, it would be analytically challenging to account for all potential spillover effects and the synergies between multiple strategies, given that the ratings for many adaptation strategies would differ substantially depending on the circumstances of their implementation.

Provide additional decision-support tools to stakeholders to help them select strategies and assemble packages of strategies. We discussed a number of trade-offs and other considerations when selecting adaptation strategies or packages of strategies that were beyond the scope of this study. We also discussed tools and resources to help aid decisionmakers. However, there are a potentially daunting number of variables to consider, including how strategies may interact with one another and whether they may produce unintended consequences, particularly at the local or regional level. A more advanced decision-support tool that includes locally or regionally relevant data could be developed and provided to decisionmakers to help them identify and assess these considerations. As a first step, we hope to publicly release the searchable strategy and decision-support tool spreadsheets developed for this project.

Factor the consequences of NCF disruption into future risk assessments. Regional disruption of an NCF can create national-level consequences, including significant threats to health and safety, economic loss, and risks to national security. CISA should consider conducting a more complete analysis of the consequences of various levels of disruption examined in our earlier report to inform the prioritization of future risk mitigation activities.

Continue to focus on developing robust communications to help stakeholders understand risk from climate change and associated adaptation strategies. Climate change–related risk presents a daunting number of communication challenges. These include uncertainty in where, when, and how specific areas will be affected by climate change; the large number of impact pathways through which risk may manifest and the number of potential adaptation strategies to address the risks; and the uncertainties, interdependencies, unintended consequences, and trade-offs inherent in these strategies. CISA should continue to invest in developing the most-robust communication tools possible to convey the risk the United States faces from climate change, its potential impact, and how this risk may be reduced through adaptation strategies, such as the ones identified in this report.

Unique Climate Adaptation Strategies Included in the Analysis

Tables A.1 through A.25 present the climate adaptation strategies we identified for the 25 NCFs that are at greatest risk from climate change effects. These strategies are targeted to the specific impact pathways considered to be most vulnerable to climate change by 2100. We list the specific adaptation strategies for each NCF (some strategies may apply to more than one impact pathway). As explained earlier in the report, a given strategy may be applicable to more than one NCF and, therefore, would appear more than once in these tables.

We also provide qualitative ratings for each of the strategies for each impact pathway considered in isolation:

- **Strength-of-evidence**: *weak*, *medium*, or *strong*. This rating includes an assessment of the extent of the existing literature and the availability of authoritative sources supporting the strategy. It refers to the strength of evidence supporting the use of the strategy to mitigate risk to the specific NCF. We considered the effectiveness rating an assessment of the extent to which the strategy, if implemented, would mitigate risk and the strength-of-evidence rating as a measure of the confidence in that rating.
- **Effectiveness**: *minor*, *moderate*, or *major impact*. This rating represents the potential effectiveness of an adaptation strategy for a given impact pathway and is based on SME assessment (and as supported by the literature) of the extent to which the strategy would be expected to reduce the risk to the NCF that each relevant impact pathway would pose if the adaptation strategy were implemented in isolation.
- **Feasibility**: *low*, *medium*, or *high*. Feasibility includes multiple aspects, such as cost; whether multiple parties needed to cooperate to implement the strategy; and whether implementing the strategy would produce an intensive need for land, skilled labor, or technologies not currently in widespread use. We also considered the compounding impact on feasibility if more than one of these factors was likely to be present.
- **Cybersecurity vulnerability**: *adds*, *does not change*, or *reduces*. This is the adaptation strategy's potential to change the level of cybersecurity risk to an NCF based on whether the strategy would make the NCF more or less vulnerable to cyber-related disruptions.

As mentioned earlier, these ratings were done for each impact pathway. When a strategy listed in the tables applied to more than one impact pathway, and the assessments differed, we used the more conservative value and indicated this with an asterisk.

TABLE A.1
Unique Climate Adaptation Strategies Included in the Analysis: Develop and Maintain Public Works and Service

Climate Adaptation Strategy	IPCC Adaptation Category	Citations	Strength of Evidence	Effectiveness	Feasibility	Cyber Vulnerability
Adopt financial mechanisms (e.g., mutual aid agreements or insurance policies)	Economic	EPA, 2017a Mount et al., 2018	Medium	Moderate	Medium	No change
Employ alternative cooling strategies (e.g., passive cooling strategies) or utilize energy-efficient equipment or appliances	Engineered and built environment	Vine, 2011 Hammer et al., 2011	Medium	Moderate	Medium	No change
Construct or strengthen local floodproofing strategies (e.g., sea walls, levees, floodgates)	Engineered and built environment	Jacob et al., 2011	Strong	Major	High	No change
Increase public awareness and/or alerts on resource consumption reduction	Informational	Vine, 2011 Hammer et al., 2011	Medium	Moderate	Medium	No change
Utilize demand management programs to effectively utilize limited supply	Government policies and programs	EPA, 2017a Mount et al., 2018	Strong	Moderate	Medium	No change
Develop resource budgets to allocate to stakeholders	Economic	EPA, 2017a Mount et al., 2018	Medium	Moderate	Medium	No change
Disseminate information to public on strategic load reduction to avoid disruption	Educational	Vine, 2011 Hammer et al., 2011	Medium	Minor	High	No change
Floodproof existing infrastructure (e.g., elevating above flood level)	Technological	Jacob et al., 2011	Strong	Major	Medium	No change
Implement price response programs to encourage behavioral consumption change	Behavioral	Vine, 2011 Hammer et al., 2011	Strong	Moderate	Medium	No change
Implement programs and/or develop workplace standards to monitor outdoor safe working conditions	Government policies and programs	Kjellstrom et al., 2019	Weak	Minor	High	No change
Improve air conditioning systems in transportation systems to accommodate extreme heat conditions	Technological	Jacob et al., 2011	Medium	Minor	Medium	No change
Incorporate requirements to provide heat illness protection measures (e.g., shading devices, cooling centers)	Laws and regulations	Kjellstrom et al., 2019	Strong	Minor	Medium	No change
Incorporate redundant or decentralized power sources	Technological	Brody, Rogers, and Siccardo, 2019	Strong	Moderate	Medium	Reduces

Table A.1—Continued

Climate Adaptation Strategy	IPCC Adaptation Category	Citations	Strength of Evidence	Effectiveness	Feasibility	Cyber Vulnerability
Increase drainage capacity to account for changing climate conditions	Engineered and built environment	Rattanachot et al., 2015	Strong	Major	High	No change
Increase storage capacity and/or modify existing infrastructure to increase resource supply	Engineered and built environment	EPA, 2017a; Mount et al., 2018	Medium	Major	Medium	No change
Invest in alternative sources of supply	Engineered and built environment	EPA, 2017a; Mount et al., 2018	Strong	Major	Medium	Reduces
Manage land/vegetation (e.g., create defensible space around infrastructure)	Ecosystem-based	Halofsky et al., 2021	Strong	Major	High	No change
Rebuild and/or maintain critical infrastructure to account for climate change	Engineered and built environment	EPA, 2017a; Mount et al., 2018	Strong	Major	Medium	No change
Reduce urban heat island effect (e.g., cool pavements or roofs, increasing vegetation abundance)	Engineered and built environment	Larsen, 2015	Strong	Moderate	High	No change
Reschedule work activities outside when there is risk of heat-related illness	Behavioral	Luber and McGeehin, 2008	Weak	Minor	Medium	No change
Improve storage and recovery of existing supply sources (e.g., groundwater)	Ecosystem-based	EPA, 2017a; Mount et al., 2018	Strong	Major	Medium	No change
Strengthen infrastructure codes and standards	Laws and regulations	Gunawansa and Kua, 2014; Federal Emergency Management Agency (FEMA), 2020; Multi-Hazard Mitigation Council, 2019	Strong	Moderate	High	No change
Use alternative materials that can withstand or adapt to climate change	Technological	Rattanachot et al., 2015; Jacob et al., 2011	Medium	Moderate	Medium	No change
Utilize less water-intensive equipment for agricultural and irrigation purposes	Technological	EPA, 2017a; Mount et al., 2018	Medium	Moderate	High	No change

TABLE A.2
Unique Climate Adaptation Strategies Included in the Analysis: Distribute Electricity

Climate Adaptation Strategy	IPCC Adaptation Category	Citation	Strength of Evidence	Effectiveness	Feasibility	Cyber Vulnerability
Develop advanced visualization and information systems	Informational	Panteli and Mancarella, 2015	Strong	Moderate	High	Adds
Employ bushfire reduction and prescribed fire	Ecosystem-based	Noble et al., 2014	Strong	Moderate	Medium	No change
Deploy distributed generation: distributed PV, microgrids, and minigrids	Technological	Gholami, Aminifar, and Shahidehpour, 2016	Strong	Minor*	Medium	Adds
Develop smart grids	Technological	Cox et al., 2017	Strong	Minor*	Medium	Adds
Elevate electrical infrastructure	Engineered and built environment	Linnenluecke, Stathakis, and Griffiths, 2011	Strong	Major	Medium	No change
Implement enhanced demand-side efficiency and reliability	Educational Behavioral Laws and regulations	Cox et al., 2017	Medium	Minor*	Medium*	No change
Implement enhanced supply-side efficiency and reliability	Technological	Stephens et al., 2013	Strong	Minor	Medium	Reduces
Establish and adopt equipment design standards	Laws and regulations	Allen-Dumas, Binita, and Cunliff, 2019	Medium	Major	Medium	No change
Add hardening measures	Engineered and built environment	Panteli and Mancarella, 2015	Strong	Moderate	Medium	No change
Implement demand-sensitive electricity tariffs	Economic	Noble et al., 2014	Strong	Major	Medium	No change
Improve climate value at risk assessment for utilities and assets	Informational	Dietz et al., 2016	Strong	Major	Medium	No change
Improve integrated electricity planning approaches	Informational	Cox et al., 2017	Medium	Major	Medium	No change*
Employ improved hazard mapping and monitoring technology	Technological	Noble et al., 2014	Medium	Moderate	Medium	Adds
Increase generation capacity reserve margins	Economic	Allen-Dumas, Binita, and Cunliff, 2019	Medium	Moderate	Medium	Adds
Perform integrated coastal zone management	Government policies and programs	Noble et al., 2014	Strong	Moderate	Medium	No change

Table A.2—Continued

Climate Adaptation Strategy	IPCC Adaptation Category	Citation	Strength of Evidence	Effectiveness	Feasibility	Cyber Vulnerability
Maintain backup components and materials	Engineered and built environment	Panteli and Mancarella, 2015	Strong	Minor	High	No change
Enhance resource efficiency	Behavioral	Stephens et al., 2013	Strong	Minor	Medium	Reduces
Leverage national and regional adaptation plans	Government policies and programs	Noble et al., 2014	Strong	Major	Low	No change
Relocate electrical infrastructure	Engineered and built environment	Linnenluecke, Stathakis, and Griffiths, 2011	Strong	Major	Medium	No change
Build sea walls and coastal protection structures	Engineered and built environment	Noble et al., 2014	Strong	Moderate	Low	No change
Strengthen utility mutual aid agreements	Services	Harrington and Cole, 2022	Strong	Moderate	High	No change
Clear transmission line rights-of-way (e.g., tree trimming, vegetation management)	Ecosystem-based	Panteli and Mancarella, 2015	Strong	Minor	High	No change
Install underground distribution conductor	Engineered and built environment	Taylor and Roald, 2021	Strong	Major	Medium	No change
Improve underground distribution conductor code	Laws and regulations	Taylor and Roald, 2021	Strong	Major	Medium	No change
Carry out vulnerability assessments	Informational	Noble et al., 2014	Strong	Minor	High	No change
Conserve and restore wetlands and floodplains	Ecosystem-based	Noble et al., 2014	Strong	Moderate	Low	No change

NOTE: When a strategy listed in the tables applied to more than one impact pathway, and the assessments differed, we used the more conservative value and indicated this with an asterisk.

TABLE A.3
Unique Climate Adaptation Strategies Included in the Analysis: Educate and Train

Climate Adaptation Strategy	IPCC Adaptation Category	Citation	Strength of Evidence	Effectiveness	Feasibility	Cyber Vulnerability
Build levees and related flood control structures around schools	Engineered and built environment	Costa et al., 2021	Strong	Moderate	High	No change
Develop sandbagging and related measures to protect against minor floods	Behavioral	Costa et al., 2021	Strong	Moderate	High	No change
Do evacuation planning	Government policies and programs	Moritz et al., 2014	Strong	Moderate	High	No change
Raise school buildings	Engineered and built environment	Costa et al., 2021	Strong	Moderate	High	No change
Require new facilities to be built outside floodplains	Laws and regulations	FEMA, 2017	Strong	Moderate	Medium	No change
Retrofit and fire-harden educational facilities	Engineered and built environment	Moritz et al., 2014	Strong	Moderate	Medium	No change
Retrofit existing buildings and enforce stronger codes for future buildings	Laws and regulations	FEMA, 2017	Strong	Moderate	High	No change
Retrofit existing buildings to harden against fire	Engineered and built environment	FEMA, 2017	Strong	Moderate	High	No change
Retrofit existing buildings to harden against wind/rain	Engineered and built environment	FEMA, 2017	Strong	Moderate	High	No change
Strengthen online and remote instruction capabilities	Technological	Costa et al., 2021	Strong	Moderate	High	Adds
Strengthen infrastructure codes and standards	Laws and regulations	FEMA, 2017	Strong	Moderate	High	No change
Use fire-resistant landscaping on school grounds	Ecosystem-based	Moritz et al., 2014	Strong	Moderate	Medium	No change

TABLE A.4

Unique Climate Adaptation Strategies Included in the Analysis: Enforce Law

Climate Adaptation Strategy	IPCC Adaptation Category	Citation	Strength of Evidence	Effectiveness	Feasibility	Cyber Vulnerability
Provide air conditioning and improve conditions at correctional facilities	Engineered and built environment	Jones, 2019	Strong	Moderate	High	No change
Perform cyber investigations and monitor federal and other assistance programs or funds	Technological	CISA, 2020 Grice, 2021 Johnson, 2016	Medium	Major	Medium	Adds
Communicate early warnings and evacuation advisories to mitigate demand surges	Technological	Gibbs and Holloway, 2013	Medium	Moderate	High	No change
Hire adequate numbers of personnel at the agency and regional levels for surges	Government policies and programs	Mostyn et al., 2019	Strong	Moderate	Medium	No change
Improve environment/environmental design (including trees and shade)	Ecosystem-based	Donovan and Prestemon, 2012 Troy, Grove, and O'Neill-Dunne, 2012	Medium	Minor	High	No change
Increase federal and state grants for hiring (e.g., Community Oriented Policing Services [COPS] program)	Government policies and programs	Roth and Ryan, 2000 Mello, 2019	Strong	Major	Medium	No change
Site public safety facilities in low-risk areas and apply construction approaches that mitigate damage	Engineered and built environment	Analyst subject-matter expertise	Weak	Moderate	Low	No change
Rehire or allow retirees to fill active positions	Government policies and programs	Harrison, 2020	Medium	Minor	High	No change

57

TABLE A.5
Unique Climate Adaptation Strategies Included in the Analysis: Exploration and Extraction of Fuels

Climate Adaptation Strategy	IPCC Adaptation Category	Citation	Strength of Evidence	Effectiveness	Feasibility	Cyber Vulnerability
Improve climate value at-risk assessment for utilities and assets	Informational	Dietz et al., 2016	Strong	Major	Medium	No change
Employ improved hazard mapping and monitoring technology	Technological	Noble et al., 2014	Medium	Moderate	Medium	Adds
Maintain backup components and materials	Engineered and built environment	Panteli and Mancarella, 2015	Strong	Minor	High	No change
Manage fuel supply chains	Informational	Busch, 2011	Medium	Moderate	Medium*	No change
Utilize and enhance municipal water management programs	Government policies and programs	Noble et al., 2014	Strong	Moderate	Medium	No change
Leverage national and regional adaptation plans	Government policies and programs	Noble et al., 2014	Strong	Major	Low	No change
Carry out vulnerability assessments	Informational	Noble et al., 2014	Strong	Minor	High	No change

NOTE: When a strategy listed in the tables applied to more than one impact pathway, and the assessments differed, we used the more conservative value and indicated this with an asterisk.

TABLE A.6

Unique Climate Adaptation Strategies Included in the Analysis: Generate Electricity

Climate Adaptation Strategy	IPCC Adaptation Category	Citation	Strength of Evidence	Effectiveness	Feasibility	Cyber Vulnerability
Develop advanced visualization and information systems	Informational	Panteli and Mancarella, 2015	Strong	Moderate	High	Adds
Employ bushfire reduction and prescribed fire	Ecosystem-based	Noble et al., 2014	Strong	Moderate	Medium	No change
Deploy distributed generation: distributed PV, microgrids, and minigrids	Technological	Gholami, Aminifar, and Shahidehpour, 2016	Strong	Moderate	Medium	Adds
Develop smart grids	Technological	Cox et al., 2017	Strong	Moderate	Medium	Adds
Elevate electrical infrastructure	Engineered and built environment	Linnenluecke, Stathakis, and Griffiths, 2011	Strong	Major	Medium	No change
Implement enhanced demand-side efficiency and reliability	Educational Behavioral Laws and regulations	Cox et al., 2017	Medium	Minor*	Medium*	No change
Implement enhanced supply-side efficiency and reliability	Technological	Stephens et al., 2013	Strong	Minor	Medium	Reduces
Add hardening measures	Engineered and built environment	Panteli and Mancarella, 2015	Strong	Moderate	Medium	No change
Implement demand-sensitive electricity tariffs	Economic	Noble et al., 2014	Strong	Major	Medium	No change
Improve climate value at risk assessment for utilities and assets	Informational	Dietz et al., 2016	Strong	Major	Medium	No change
Improve integrated electricity planning approaches	Informational	Cox et al., 2017	Medium	Major	Medium	No change*
Employ improved hazard mapping and monitoring technology	Technological	Noble et al., 2014	Medium	Moderate	Medium	Adds

Table A.6—Continued

Climate Adaptation Strategy	IPCC Adaptation Category	Citation	Strength of Evidence	Effectiveness	Feasibility	Cyber Vulnerability
Perform integrated coastal zone management	Government policies and programs	Noble et al., 2014	Strong	Moderate	Medium	No change
Maintain backup components and materials	Engineered and built environment	Panteli and Mancarella, 2015	Strong	Minor	High	No change
Manage supply chains	Informational	Macknick et al., 2012	Medium	Moderate	High	No change
Enhance resource efficiency	Behavioral	Stephens et al., 2013	Strong	Minor	Medium	Reduces
Utilize and enhance municipal water management programs	Government policies and programs	Noble et al., 2014	Strong	Moderate	Medium	No change
Leverage national and regional adaptation plans	Government policies and programs	Noble et al., 2014	Strong	Major	Low	No change
Relocate electrical infrastructure	Engineered and built environment	Linnenluecke, Stathakis, and Griffiths, 2011	Strong	Major	Medium	No change
Substitute renewable energy technology	Technological	Noble et al., 2014 Macknick et al., 2012	Strong	Major	High	Adds
Build sea walls and coastal protection structures	Engineered and built environment	Noble et al., 2014	Strong	Moderate	Low	No change
Strengthen utility mutual aid agreements	Services	Harrington and Cole, 2022	Strong	Moderate	High	No change
Carry out vulnerability assessments	Informational	Noble et al., 2014	Strong	Minor	High	No change
Conserve and restore wetlands and floodplains	Ecosystem-based	Noble et al., 2014	Strong	Moderate	Low	No change

NOTE: When a strategy listed in the tables applied to more than one impact pathway, and the assessments differed, we used the more conservative value and indicated this with an asterisk.

TABLE A.7

Unique Climate Adaptation Strategies Included in the Analysis: Maintain Supply Chains

Climate Adaptation Strategy	IPCC Adaptation Category	Citation	Strength of Evidence	Effectiveness	Feasibility	Cyber Vulnerability
Employ blockchain technology to increase supply chain transparency	Technological	Etemadi et al., 2021; Advisen, 2013; Baskin, 2020	Medium	Minor*	Medium	Reduces
Construct and reinforce dikes and embankments	Engineered and built environment	Ward et al., 2017; Jongman, 2018; Krausmann and Mushtaq, 2008; Organisation for Economic Co-operation and Development (OECD), 2018; Doll et al., 2011	Strong	Moderate	Low	No change
Dredge riverbeds to increase capacity	Engineered and built environment	Doll et al., 2011	Medium	Moderate	Medium	No change
Improve port substitution and redundancy	Engineered and built environment	Verschuur, Koks, and Hall, 2020	Weak	Minor	Low	Reduces
Lengthen runways	Engineered and built environment	Arent et al., 2014	Medium	Minor	Medium	No change
Minimize or remove hard surfaces and replace with natural infrastructure	Engineered and built environment	Doll et al., 2011	Medium	Moderate	Medium	No change
Shift away from air transport to other modes	Informational	Doll et al., 2011	Weak	Minor	Medium	No change
Carry out port mapping and planning	Informational	Rice, Trepte, and Cottrill, 2013	Weak	Minor	High	Reduces
Raise docks and wharf levels	Engineered and built environment	Doll et al., 2011	Medium	Moderate	Low	No change
Build sea walls and coastal protection structures	Engineered and built environment	Han and Mozumder, 2021; OECD, 2018; Nicholls and Tol, 2006; Han et al., 2020; Doll et al., 2011; Peacock and Husein, 2011	Strong	Moderate	Low	No change
Employ smart technologies for detecting abnormal events	Technological	Doll et al., 2011	Medium	Moderate	Medium	Reduces

Table A.7—Continued

Climate Adaptation Strategy	IPCC Adaptation Category	Citation	Strength of Evidence	Effectiveness	Feasibility	Cyber Vulnerability
Stockpile critical goods	Laws and regulations	Goentzel and Windle, 2017 Paul and MacDonald, 2016 Pan et al., 2020 Paul and Hariharan, 2012	Strong	Moderate	High	No change
Make technological improvements to airplanes	Technological	Arent et al., 2014	Weak	Moderate	Medium	No change
Enhance zoning and land use and relocate critical assets in vulnerable areas	Laws and regulations	Romero-Lankao et al., 2014 Baskin, 2020 OECD, 2018 Doll et al., 2011	Strong	Major	Low	No change

NOTE: When a strategy listed in the tables applied to more than one impact pathway, and the assessments differed, we used the more conservative value and indicated this with an asterisk.

TABLE A.8
Unique Climate Adaptation Strategies Included in the Analysis: Manage Hazardous Materials

Climate Adaptation Strategy	IPCC Adaptation Category	Citation	Strength of Evidence	Effectiveness	Feasibility	Cyber Vulnerability
Install building insulation and mechanical cooling	Technological	Baxter et al., 2013; Cutter et al., 2012; Pachon, Dailey, and Treimel, 2014; United Nations Environment Programme (UNEP) et al., 2021; EPA, 2016; EPA, 2014a	Strong	Moderate	High	No change
Invest in and implement disaster preparedness and response	Government policies and programs	Baxter et al., 2013; Cutter et al., 2012; Pachon, Dailey, and Treimel, 2014; UNEP et al., 2021; EPA, 2016; EPA, 2014a	Medium*	Minor*	Medium*	No change*
Implement early warning systems	Technological	Baxter et al., 2013; Cutter et al., 2012; Pachon, Dailey, and Treimel, 2014; UNEP et al., 2021; EPA, 2016; EPA, 2014a	Weak*	Moderate*	Medium*	Adds
Implement natural resource management to leverage ecosystem services and employ adaptive land-use management	Ecosystem-based	Baxter et al., 2013; Cutter et al., 2012; Pachon, Dailey, and Treimel, 2014; UNEP et al., 2021; EPA, 2016; EPA, 2014a	Strong	Major	High	Reduces
Build flood levees, culverts, and stormwater management structures	Engineered and built environment	Baxter et al., 2013; Cutter et al., 2012; Pachon, Dailey, and Treimel, 2014; UNEP et al., 2021; EPA, 2016; EPA, 2014a	Strong	Moderate*	High	No change*
Employ hazard mapping and monitoring technology	Technological	Baxter et al., 2013; Cutter et al., 2012; Pachon, Dailey, and Treimel, 2014; UNEP et al., 2021; EPA, 2016; EPA, 2014a	Strong	Moderate	High	No change

Table A.8—Continued

Climate Adaptation Strategy	IPCC Adaptation Category	Citation	Strength of Evidence	Effectiveness	Feasibility	Cyber Vulnerability
Enhance insurance and related products	Economic	Baxter et al., 2013 Cutter et al., 2012 Pachon, Dailey, and Treimel, 2014 UNEP et al., 2021 EPA, 2016 EPA, 2014a	Strong	Moderate*	High	No change
Strengthen municipal services including fire/hazmat response, water, and sanitation	Services	Baxter et al., 2013 Cutter et al., 2012 Pachon, Dailey, and Treimel, 2014 UNEP et al., 2021 EPA, 2016 EPA, 2014a	Medium*	Moderate*	High	No change
Carry out scenario planning and vulnerability assessments	Informational	Baxter et al., 2013 Cutter et al., 2012 Pachon, Dailey, and Treimel, 2014 UNEP et al., 2021 EPA, 2016 EPA, 2014a	Medium*	Minor	Low*	Reduces
Build sea walls and coastal protection structures	Engineered and built environment	Baxter et al., 2013 Cutter et al., 2012 Pachon, Dailey, and Treimel, 2014 UNEP et al., 2021 EPA, 2016 EPA, 2014a	Strong	Moderate*	High	No change
Strengthen infrastructure codes and standards for containment structures	Engineered and built environment	Baxter et al., 2013 Cutter et al., 2012 Pachon, Dailey, and Treimel, 2014 UNEP et al., 2021 EPA, 2016 EPA, 2014a	Strong	Major	High	Reduces
Enhance zoning and inspections	Laws and regulations	Baxter et al., 2013 Cutter et al., 2012 Pachon, Dailey, and Treimel, 2014 UNEP et al., 2021 EPA, 2016 EPA, 2014a	Medium*	Minor*	Low*	No change*

NOTE: When a strategy listed in the tables applied to more than one impact pathway, and the assessments differed, we used the more conservative value and indicated this with an asterisk.

TABLE A.9

Unique Climate Adaptation Strategies Included in the Analysis: Manage Wastewater

Climate Adaptation Strategy	IPCC Adaptation Category	Citation	Strength of Evidence	Effectiveness	Feasibility	Cyber Vulnerability
Build flood barriers to protect infrastructure or relocate facilities to higher elevations	Engineered and built environment	EPA, undated c	Strong	Moderate	Medium	No change
Implement policies and procedures for post-fire repairs	Government policies and programs	EPA, undated c	Strong	Minor	Medium	No change
Implement policies and procedures for post-flood repairs	Government policies and programs	EPA, undated c	Strong	Minor	Medium	No change
Increase capacity for wastewater and stormwater collection, treatment, and storage	Engineered and built environment	EPA, undated c	Strong	Major	Medium	No change
Model and reduce inflow/infiltration in the sewer system	Technological	EPA, undated c	Strong	Major	Medium	No change
Plan and establish alternative or on-site power supply	Engineered and built environment	EPA, undated c	Strong	Moderate	Medium	Adds*
Study response of nearby wetlands to storm surge events	Technological	EPA, undated c	Strong	Major	High	No change
Update fire models and practice fire management plans	Government policies and programs	EPA, undated c	Strong	Moderate	High	No change

NOTE: When a strategy listed in the tables applied to more than one impact pathway, and the assessments differed, we used the more conservative value and indicated this with an asterisk.

TABLE A.10
Unique Climate Adaptation Strategies Included in the Analysis: Manufacture Equipment

Climate Adaptation Strategy	IPCC Adaptation Category	Citation	Strength of Evidence	Effectiveness	Feasibility	Cyber Vulnerability
Employ blockchain technology to increase supply chain transparency	Technological	Etemadi et al., 2021	Medium	Minor	Medium	Reduces
Develop workplace heat standards	Laws and regulations	Occupational Safety and Health Administration (OSHA), 2021; White House, 2021; Rowlinson and Jia, 2014	Strong	Moderate	High	No change
Implement early warning systems and educational outreach	Informational	Kjellstrom et al., 2019; Nabeel et al., 2021	Medium	Minor	Low	No change
Install energy-efficient cooling systems	Technological	Romero-Lankao et al., 2014	Medium	Minor	Medium	No change
Harden facilities and use stronger design standards	Engineered and built environment	Qin, Khakzad, and Zhu, 2020; Romero-Lankao et al., 2014; OECD, 2018; Arent et al., 2014	Strong	Major	Medium	No change
Build levees and dikes	Engineered and built environment	Ward et al., 2017; Jongman, 2018; Krausmann and Mushtaq, 2008; OECD, 2018; Peacock and Husein, 2011	Strong	Moderate	Low	No change
Carry out preevent planning and training programs, guidelines, and best practices	Informational	Arent et al., 2014; Kim and Bui, 2019; Leonard, 2019; National Integrated Drought Information System, undated	Medium*	Moderate	High	Reduces
Build sea walls and coastal protection structures	Engineered and built environment	Han and Mozumder, 2021; OECD, 2018; Nicholls and Tol, 2006; Han et al., 2020; Peacock and Husein, 2011	Strong	Moderate	Low	No change
Stockpile critical goods	Laws and regulations	Goentzel and Windle, 2017; Paul and MacDonald, 2016; Pan et al., 2020; Paul and Hariharan, 2012	Strong	Moderate	High	No change

Table A.10—Continued

Climate Adaptation Strategy	IPCC Adaptation Category	Citation	Strength of Evidence	Effectiveness	Feasibility	Cyber Vulnerability
Map supply chains, track inputs, and diversify suppliers	Informational	Baskin, 2020 Pettit, Croxton, and Fiksel, 2013	Medium	Moderate	Medium	Reduces
Enhance zoning and land use and relocate critical assets in vulnerable areas	Laws and regulations	Romero-Lankao et al., 2014 Baskin, 2020 OECD, 2018	Strong	Major	Low	No change

NOTE: When a strategy listed in the tables applied to more than one impact pathway, and the assessments differed, we used the more conservative value and indicated this with an asterisk.

TABLE A.11
Unique Climate Adaptation Strategies Included in the Analysis: Prepare for and Manage Emergencies

Climate Adaptation Strategy	IPCC Adaptation Category	Citation	Strength of Evidence	Effectiveness	Feasibility	Cyber Vulnerability
Build a volunteer reserve corps	Services	Frasca, 2010; U.S. Fire Administration, 2008; Ready.gov, undated	Medium	Minor	High	No change
Develop and exercise mutual aid agreements to cope with demand surge during disasters	Services	Harrington and Cole, 2022 National Emergency Management Association, undated	Strong	Major	High	No change
Develop and exercise plans for relocation and repositioning when weather events threaten facilities	Behavioral	U.S. Fire Administration, 2008	Strong	Moderate	High	No change
Increase workforce	Services	U.S. Government Accountability Office, 2020 Frank, 2019	Strong	Major	High	No change

TABLE A.12

Unique Climate Adaptation Strategies Included in the Analysis: Produce and Provide Agricultural Products and Services

Climate Adaptation Strategy	IPCC Adaptation Category	Citation	Strength of Evidence	Effectiveness	Feasibility	Cyber Vulnerability
Support changes in producers' practices and patterns of activity	Behavioral	Cai et al., 2015	Strong	Minor	Low	No change
Implement ecological restoration, preservation, conservation, and natural resource management	Ecosystem-based	Alexandratos et al., 2019 Álvarez-Berríos et al., 2021 Bravo, Leiras, and Oliveira, 2020 California Climate and Agriculture Network, undated Charnley et al., 2015 FEMA, 2021 Johnson et al., 2020 King et al., 2021 Meerow and Keith, 2021 Schulz, 2017 Parker et al., 2020 Tully et al., 2019 Janowiak et al., 2016 U.S. Department of Agriculture (USDA), 2021a Wiener, Álvarez-Berríos, and Lindsey, 2020	Weak*	Minor*	Low*	No change
Invest in infrastructure to support key (water) resources	Economic	Cai et al., 2015	Medium	Moderate	Low	No change

Table A.12—Continued

Climate Adaptation Strategy	IPCC Adaptation Category	Citation	Strength of Evidence	Effectiveness	Feasibility	Cyber Vulnerability
Improve government planning, preparedness, and/or investment	Government policies and programs	Álvarez-Berríos et al., 2021 Bressers, Bressers, and Larrue, 2016 American Farm Bureau Federation, 2021 FEMA, 2021 Fu et al., 2013 Jackson and Rosenberg, 2010 King et al., 2021 McGranahan et al., 2013 Meerow and Keith, 2021 OECD, 2021 Pal, Patel, and Banik, 2021 Parker et al., 2020 Tigchelaar, Battisti, and Spector, 2020 Turek-Hankins et al., 2021 USDA, 2021a USDA, 2021b Wickham et al., 2019 Wildland Fire Leadership Council, undated Charnley et al., 2015	Weak*	Minor*	Low*	Adds*
Implement hazard and vulnerability mapping and early warning and response systems	Informational	Fu et al., 2013 Milly et al., 2008 Svoboda et al., 2015 University of Nebraska–Lincoln, undated Wickham et al., 2019	Strong	Minor	Medium*	No change
Implement technologies to monitor and mitigate climate impacts	Technological	Alexandratos et al., 2019 Álvarez-Berríos et al., 2021 Bravo, Leiras, and Oliveira, 2020 Giller et al., 2015 Lafferty et al., 2021 Meerow and Keith, 2021 OECD, 2021 Parker et al., 2020 Reams et al., 2005 Shirzaei et al., 2021 Svoboda et al., 2015 University of Nebraska–Lincoln, undated USDA, 2021a	Weak*	Minor*	Low*	Adds

Table A.12—Continued

Climate Adaptation Strategy	IPCC Adaptation Category	Citation	Strength of Evidence	Effectiveness	Feasibility	Cyber Vulnerability
Adapt infrastructure to manage and store water resources	Engineered and built environment	Alexandratos et al., 2019 Bravo, Leiras, and Oliveira, 2020 USDA, 2021a Janowiak et al., 2016	Strong	Moderate	Medium*	No change
Adapt infrastructure to manage and store land resources	Engineered and built environment	Álvarez-Berríos et al., 2021 Bravo, Leiras, and Oliveira, 2020 FEMA, 2021 Meerow and Keith, 2021 Schulz, 2017 Parker et al., 2020 Tully et al., 2019 USDA, 2021a Janowiak et al., 2016 Wiener, Álvarez-Berríos, and Lindsey, 2020	Weak*	Minor*	Low*	No change
Enhance provision and uptake of disaster assistance and support services	Services	Álvarez-Berríos et al., 2021 Charnley et al., 2015 Schulz, 2017 OECD, 2021 Taylor, 2021 USDA, 2021a USDA, undated b USDA, 2021b EPA, undated a Wiener, Álvarez-Berríos, and Lindsey, 2020 Wildland Fire Leadership Council, undated	Weak	Minor*	Medium*	No change
Enhance provision and uptake of insurance services	Economic	Deryugina and Konar, 2017 Kane et al., 2021 Maestro, Garrido, and Bielza, 2018	Weak	Minor	Low*	No change

Table A.12—Continued

Climate Adaptation Strategy	IPCC Adaptation Category	Citation	Strength of Evidence	Effectiveness	Feasibility	Cyber Vulnerability
Enact regulations for key resources and laws that support risk reduction	Laws and regulations	Bressers, Bressers, and Larrue, 2016 California Climate and Agriculture Network, undated Charnley et al., 2015 Fu et al., 2013 Gopalakrishnan et al., 2019 Jackson and Rosenberg, 2010 King et al., 2021 Marmet, 2013 McGranahan et al., 2013 Pal, Patel, and Banik, 2021 Tigchelaar, Battisti, and Spector, 2020 Wickham et al., 2019	Weak*	Minor*	Low*	No change
Educate agricultural workforce and raise awareness for agricultural producers	Educational	Álvarez-Berríos et al., 2021 Culp et al., 2011 EPA, undated a Jackson and Rosenberg, 2010 National Hog Farmer, 2017 OECD, 2021 Pal, Patel, and Banik, 2021 Svoboda et al., 2015 University of Nebraska–Lincoln, undated USDA, undated b Wiener, Álvarez-Berríos, and Lindsey, 2020 USDA, 2021a	Weak	Minor	Medium*	No change

NOTE: When a strategy listed in the tables applied to more than one impact pathway, and the assessments differed, we used the more conservative value and indicated this with an asterisk.

TABLE A.13

Unique Climate Adaptation Strategies Included in the Analysis: Produce Chemicals

Climate Adaptation Strategy	IPCC Adaptation Category	Citation	Strength of Evidence	Effectiveness	Feasibility	Cyber Vulnerability
Employ blockchain technology to increase supply chain transparency	Technological	Etemadi et al., 2021	Medium	Minor	Medium	Reduces
Develop workplace heat standards	Laws and regulations	OSHA, 2021 White House, 2021 Rowlinson and Jia, 2014	Strong	Moderate	High	No change
Harden facilities and use stronger design standards	Engineered and built environment	Qin, Khakzad, and Zhu, 2020 Romero-Lankao et al., 2014 OECD, 2018 Arent et al., 2014	Strong	Major	Medium	No change
Build levees and dikes	Engineered and built environment	Ward et al., 2017 Jongman, 2018 Krausmann and Mushtaq, 2008 OECD, 2018 Peacock and Husein, 2011	Strong	Moderate	Low	No change
Carry out preevent planning and training programs, guidelines, and best practices	Informational	Krausmann and Mushtaq, 2008 Reniers et al., 2018 Leonard, 2019 Kim and Bui, 2019 Arent et al., 2014	Strong	Moderate	High	Reduces
Build sea walls and coastal protection structures	Engineered and built environment	Han and Mozumder, 2021 OECD, 2018 Nicholls and Tol, 2006 Han et al., 2020 Peacock and Husein, 2011	Strong	Moderate	Low	No change
Stockpile critical goods	Laws and regulations	Goentzel and Windle, 2017 Paul and MacDonald, 2016 Pan et al. 2020 Paul and Hariharan, 2012	Strong	Moderate	High	No change
Map supply chains, track inputs, and diversify suppliers	Informational	Advisen, 2013 Baskin, 2020 Pettit, Croxton, and Fiksel, 2013	Strong	Moderate	Medium	Reduces

Table A.13—Continued

Climate Adaptation Strategy	IPCC Adaptation Category	Citation	Strength of Evidence	Effectiveness	Feasibility	Cyber Vulnerability
Enhance zoning and land use and relocate critical assets in vulnerable areas	Laws and regulations	Reniers et al., 2018 Romero-Lankao et al., 2014 Baskin, 2020 OECD, 2018	Strong	Major	Low	No change

TABLE A.14
Unique Climate Adaptation Strategies Included in the Analysis: Provide Housing

Climate Adaptation Strategy	IPCC Adaptation Category	Citation	Strength of Evidence	Effectiveness	Feasibility	Cyber Vulnerability
Promote hazard awareness, traditional knowledge-sharing, and community surveys	Educational	Brugmann, 2011 Extreme Weather Response Task Force, 2021 Hurd et al., 2017 Martin et al., 2013 HUD, 2015 Stagrum et al., 2020 UNEP, 2021 HUD, 2021 HUD, 2014 HUD, 2022	Strong	Moderate	High	No change
Establish building codes and standards	Engineered and built environment	Brugmann, 2011 Extreme Weather Response Task Force, 2021 Hurd et al., 2017 Martin et al., 2013 HUD, 2015 Stagrum et al., 2020 UNEP, 2021 HUD, 2021 HUD, 2014 HUD, 2022	Strong	Major	High	No change
Install building insulation and mechanical cooling	Technological	Brugmann, 2011 Extreme Weather Response Task Force, 2021 Hurd et al., 2017 Martin et al., 2013 HUD, 2015 Stagrum et al., 2020 UNEP, 2021 HUD, 2021 HUD, 2014 HUD, 2022	Strong	Major	High	No change

Table A.14—Continued

Climate Adaptation Strategy	IPCC Adaptation Category	Citation	Strength of Evidence	Effectiveness	Feasibility	Cyber Vulnerability
Raise awareness through media communications	Educational	Brugmann, 2011 Extreme Weather Response Task Force, 2021 Hurd et al., 2017 Martin et al., 2013 HUD, 2015 Stagrum et al., 2020 UNEP, 2021 HUD, 2021 HUD, 2014 HUD, 2022	Strong	Major	High	No change
Invest and implement disaster preparedness and response	Government policies and programs	Brugmann, 2011 Extreme Weather Response Task Force, 2021 Martin et al., 2013 HUD, 2015 Stagrum et al., 2020 UNEP, 2021 HUD, 2021 HUD, 2014 HUD, 2022	Strong	Moderate*	High	No change
Implement early warning systems	Technological Engineered and built environment	Brugmann, 2011 Extreme Weather Response Task Force, 2021 Hurd et al., 2017 Martin et al., 2013 HUD, 2015 Stagrum et al., 2020 UNEP, 2021 HUD, 2021 HUD, 2014 HUD, 2022	Strong	Major	High	No change

Table A.14—Continued

Climate Adaptation Strategy	IPCC Adaptation Category	Citation	Strength of Evidence	Effectiveness	Feasibility	Cyber Vulnerability
Implement natural resource management to leverage ecosystem services and employ adaptive land-use management	Ecosystem-based	Brugmann, 2011 Extreme Weather Response Task Force, 2021 Hurd et al., 2017 Martin et al., 2013 HUD, 2015 Stagrum et al., 2020 UNEP, 2021 HUD, 2021 HUD, 2014 HUD, 2022	Strong	Major	High	No change
Conduct evacuation planning	Behavioral	Brugmann, 2011 Extreme Weather Response Task Force, 2021 Hurd et al., 2017 Martin et al., 2013 HUD, 2015 Stagrum et al., 2020 UNEP, 2021 HUD, 2021 HUD, 2014 HUD, 2022	Strong	Major	High	No change
Build flood and cyclone shelters	Engineered and built environment	Brugmann, 2011 Extreme Weather Response Task Force, 2021 Hurd et al., 2017 Martin et al., 2013 HUD, 2015 Stagrum et al., 2020 UNEP, 2021 HUD, 2021 HUD, 2014 HUD, 2022	Strong	Major	High	No change

Table A.14—Continued

Climate Adaptation Strategy	IPCC Adaptation Category	Citation	Strength of Evidence	Effectiveness	Feasibility	Cyber Vulnerability
Build flood levees, culverts, and stormwater management structures	Engineered and built environment	Brugmann, 2011 Extreme Weather Response Task Force, 2021 Hurd et al., 2017 Martin et al., 2013 HUD, 2015 Stagrum et al., 2020 UNEP, 2021 HUD, 2021 HUD, 2014 HUD, 2022	Strong	Major	High	No change
Develop green infrastructure	Ecosystem-based	Brugmann, 2011 Extreme Weather Response Task Force, 2021 Hurd et al., 2017 Martin et al., 2013 HUD, 2015 Stagrum et al., 2020 UNEP, 2021 HUD, 2021 HUD, 2014 HUD, 2022	Strong	Moderate*	High	No change
Employ hazard mapping and monitoring technology	Technological	Brugmann, 2011 Extreme Weather Response Task Force, 2021 Hurd et al., 2017 Martin et al., 2013 HUD, 2015 Stagrum et al., 2020 UNEP, 2021 HUD, 2021 HUD, 2014 HUD, 2022	Strong	Moderate	Medium*	No change

Table A.14—Continued

Climate Adaptation Strategy	IPCC Adaptation Category	Citation	Strength of Evidence	Effectiveness	Feasibility	Cyber Vulnerability
Elevate, floodproof, or float houses	Engineered and built environment	Brugmann, 2011 Extreme Weather Response Task Force, 2021 Hurd et al., 2017 Martin et al., 2013 HUD, 2015 Stagrum et al., 2020 UNEP, 2021 HUD, 2021 HUD, 2014 HUD, 2022	Strong	Major	Medium	No change
Support household preparation	Behavioral	Brugmann, 2011 Extreme Weather Response Task Force, 2021 Hurd et al., 2017 Martin et al., 2013 HUD, 2015 Stagrum et al., 2020 UNEP, 2021 HUD, 2021 HUD, 2014 HUD, 2022	Strong	Minor*	High	No change
Enhance insurance and related products	Economic	Brugmann, 2011 Extreme Weather Response Task Force, 2021 Hurd et al., 2017 Martin et al., 2013 HUD, 2015 Stagrum et al., 2020 UNEP, 2021 HUD, 2021 HUD, 2014 HUD, 2022	Medium*	Minor*	Medium*	No change

Table A.14—Continued

Climate Adaptation Strategy	IPCC Adaptation Category	Citation	Strength of Evidence	Effectiveness	Feasibility	Cyber Vulnerability
Enhance microfinance and cash transfers	Economic	Brugmann, 2011 Extreme Weather Response Task Force, 2021 Hurd et al., 2017 Martin et al., 2013 HUD, 2015 Stagrum et al., 2020 UNEP, 2021 HUD, 2021 HUD, 2014 HUD, 2022	Strong	Major	High	No change
Increase reliance on social networks	Behavioral	Brugmann, 2011 Extreme Weather Response Task Force, 2021 Hurd et al., 2017 Martin et al., 2013 HUD, 2015 Stagrum et al., 2020 UNEP, 2021 HUD, 2021 HUD, 2014 HUD, 2022	Strong	Major	High	No change
Carry out scenario planning, vulnerability assessments, and community-based adaptation actions	Informational	Brugmann, 2011 Extreme Weather Response Task Force, 2021 Hurd et al., 2017 Martin et al., 2013 HUD, 2015 Stagrum et al., 2020 UNEP, 2021 HUD, 2021 HUD, 2014 HUD, 2022	Strong	Minor*	Medium*	No change

Table A.14—Continued

Climate Adaptation Strategy	IPCC Adaptation Category	Citation	Strength of Evidence	Effectiveness	Feasibility	Cyber Vulnerability
Build sea walls and coastal protection structures	Engineered and built environment	Brugmann, 2011 Extreme Weather Response Task Force, 22021 Hurd et al., 2017 Martin et al., 2013 HUD, 2015 Stagrum et al., 2020 UNEP, 2021 HUD, 2021 HUD, 2014 HUD, 2022	Strong	Major	High	No change
Enhance social safety nets and protections	Services	Brugmann, 2011 Extreme Weather Response Task Force, 2021 Hurd et al., 2017 Martin et al., 2013 HUD, 2015 Stagrum et al., 2020 UNEP, 2021 HUD, 2021 HUD, 2014 HUD, 2022	Strong	Moderate*	Medium*	No change
Update heating and cooling utility frameworks	Laws and regulations	Brugmann, 2011 Extreme Weather Response Task Force, 2021 Hurd et al., 2017 Martin et al., 2013 HUD, 2015 Stagrum et al., 2020 UNEP, 2021 HUD, 2021 HUD, 2014 HUD, 2022	Strong	Major	High	No change

Table A.14—Continued

Climate Adaptation Strategy	IPCC Adaptation Category	Citation	Strength of Evidence	Effectiveness	Feasibility	Cyber Vulnerability
Update shelter standards	Laws and regulations	Brugmann, 2011 Extreme Weather Response Task Force, 22021 Hurd et al., 2017 Martin et al., 2013 HUD, 2015 Stagrum et al., 2020 UNEP, 2021 HUD, 2021 HUD, 2014 HUD, 2022	Strong	Major	High	No change
Enhance zoning and inspections	Laws and regulations	Brugmann, 2011 Extreme Weather Response Task Force, 2021 Hurd et al., 2017 Martin et al., 2013 HUD, 2015 Stagrum et al., 2020 UNEP, 2021 HUD, 2021 HUD, 2014 HUD, 2022	Strong	Moderate*	High	No change

NOTE: When a strategy listed in the tables applied to more than one impact pathway, and the assessments differed, we used the more conservative value and indicated this with an asterisk.

TABLE A.15
Unique Climate Adaptation Strategies Included in the Analysis: Provide and Maintain Infrastructure

Climate Adaptation Strategy	IPCC Adaptation Category	Citation	Strength of Evidence	Effectiveness	Feasibility	Cyber Vulnerability
Adjust wind-loading standards for electricity grids	Laws and regulations	Arent et al., 2014	Medium	Minor	Medium	No change
Adopt financial mechanisms (e.g., mutual aid agreements or insurance policies)	Economic	EPA, 2017a Mount et al., 2018	Medium	Moderate	Medium	No change
Develop workplace heat standards	Laws and regulations	OSHA, 2021 White House, 2021 Rowlinson and Jia, 2014	Strong	Moderate	High	No change
Implement early warning systems and educational outreach	Informational	Kjellstrom et al., 2019 Nabeel et al., 2021 Peacock and Husein, 2011	Medium	Minor	Low	No change
Perform energy reserve planning and demand-side management	Informational	Panteli and Mancarella, 2015	Medium	Moderate	Medium	Reduces
Floodproof existing infrastructure (e.g., elevating above flood elevation)	Technological	Jacob et al., 2011	Strong	Major	Medium	No change
Implement programs and/or develop workplace standards to monitor outdoor safe working conditions	Government policies and programs	Kjellstrom et al., 2019	Weak	Minor	High	No change
Increase public hurricane shelter capacity	Services	Baker et al., 2008 Peacock and Husein, 2011	Medium	Moderate	Medium	No change
Incorporate redundant or decentralized power sources	Technological	Brody, Rogers, and Siccardo, 2019	Strong	Moderate	Medium	Reduces
Build levees and dikes	Engineered and built environment	Ward et al., 2017 Jongman, 2018 Krausmann and Mushtaq, 2008 OECD, 2018 Peacock and Husein, 2011	Strong	Moderate	Low	No change

Table A.15—Continued

Climate Adaptation Strategy	IPCC Adaptation Category	Citation	Strength of Evidence	Effectiveness	Feasibility	Cyber Vulnerability
Install microgrids to support local energy needs	Technological	Kushner, Ratner, and Schlegelmilch, 2021 U.S. Department of Energy, 2021	Medium	Moderate	Low	Reduces
Utilize natural infrastructure (e.g., mangroves)	Ecosystem-based	Jongman, 2018 Arent et al., 2014	Strong	Moderate	High	No change
Utilize natural infrastructure (e.g., marshes, wetlands)	Ecosystem-based	Jongman, 2018 OECD, 2018 Arent et al., 2014	Strong	Moderate	High	No change
Raise bridges	Engineered and built environment	Arent et al., 2014	Medium	Minor	Medium	No change
Relocate critical assets	Engineered and built environment	OECD, 2018 Arent et al., 2014	Strong	Major	Low	No change
Build sea walls and coastal protection structures	Engineered and built environment	Han and Mozumder, 2021 OECD, 2018 Nicholls and Tol, 2006 Han et al., 2020 Peacock and Husein, 2011	Strong	Moderate	Low	No change
Strengthen infrastructure codes and standards	Laws and regulations	Gunawansa and Kua, 2014 FEMA, 2020 Multi-Hazard Mitigation Council, 2019	Strong	Moderate	High	No change
Use alternative materials that can withstand or adapt to climate change	Technological	Rattanachot et al., 2015 Jacob et al., 2011	Medium	Moderate*	Medium	No change

NOTE: When a strategy listed in the tables applied to more than one impact pathway, and the assessments differed, we used the more conservative value and indicated this with an asterisk.

TABLE A.16

Unique Climate Adaptation Strategies Included in the Analysis: Provide Insurance Services

Climate Adaptation Strategy	IPCC Adaptation Category	Citation	Strength of Evidence	Effectiveness	Feasibility	Cyber Vulnerability
Implement community wildfire protection planning	Informational	Insurance Information Institute, 2019	Medium	Minor	High	No change
Harden buildings and infrastructure against fire	Engineered and built environment	U.S. Forest Service, 2021	Strong	Moderate	High	No change
Improve and measure ecological forest management	Ecosystem-based	U.S. Forest Service, 2021	Weak	Moderate	Medium	No change
Improve estimates of risk of flooding (e.g., National Flood Insurance Program Risk Rating 2.0)	Government policies and programs	American Flood Coalition, 2021 U.S. Government Accountability Office, 2021 Scata, 2022	Medium	Moderate	High	No change
Improve estimates of risk of wildfire	Informational	Irfan, 2021	Medium	Moderate	Medium	No change
Provide government reinsurance in catastrophic events	Government policies and programs	Insurance Information Institute, 2019	Medium	Moderate	Medium	No change
Strengthen building codes and standards	Laws and regulations	U.S. Forest Service, 2021	Strong	Moderate	High	No change

TABLE A.17
Unique Climate Adaptation Strategies Included in the Analysis: Provide Medical Care

Climate Adaptation Strategy	IPCC Adaptation Category	Citation	Strength of Evidence	Effectiveness	Feasibility	Cyber Vulnerability
Build a volunteer reserve corps	Services	Frasca, 2010 U.S. Fire Administration, 2008 Ready.gov, undated	Strong	Minor	High	No change
Build levees or floodwalls	Engineered and built environment	U.S. Department of Health and Human Services (HHS), 2014 FEMA, 2007	Strong	Major	High	No change
Design ventilation systems to recirculate and filter air	Engineered and built environment	NOAA, 2017 HHS, 2020	Strong	Major	High	No change
Develop and exercise mutual aid agreements to cope with demand surge during disasters	Services	Harrington and Cole, 2022 National Emergency Management Association, undated	Strong	Major	High	No change
Develop and practice shelter-in-place and evacuation plans	Behavioral	HHS, 2014 FEMA, 2007 NOAA, 2017 HHS, 2020	Strong	Major	High	No change
Elevate buildings or key components, including utilities and backups	Engineered and built environment	HHS, 2014 FEMA, 2007	Strong	Major	High	No change
Implement fire safety building features	Engineered and built environment	Joint Commission, undated NOAA, 2017 HHS, 2020	Strong	Major	High	No change
Increase workforce	Services	National Academies of Sciences, Engineering, and Medicine, 2021 American Hospital Association, 2021 Mercer, 2021 Lasater et al., 2021	Strong	Major	High	No change
Maintain backup systems for power, water, and medical records	Engineered and built environment	HHS, 2014 FEMA, 2007 NOAA, 2017 HHS, 2020	Strong	Major	High	No change
Prohibit facilities from being built in 500-year flood zones	Laws and regulations	HHS, 2014 FEMA, 2007	Strong	Major	High	No change
Prohibit facilities from being built in very high fire severity zones	Engineered and built environment	Adelaine et al., 2017	Strong	Major	Medium	No change

Table A.17—Continued

Climate Adaptation Strategy	IPCC Adaptation Category	Citation	Strength of Evidence	Effectiveness	Feasibility	Cyber Vulnerability
Reinforce and floodproof structures, doors, windows, and vents	Engineered and built environment	HHS, 2014 FEMA, 2007	Strong	Major	High	No change
Relocate facilities away from flood areas	Engineered and built environment	HHS, 2014 FEMA, 2007	Strong	Major	Medium	No change
Relocate facilities away from very high fire severity zones	Engineered and built environment	Adelaine et al., 2017	Strong	Major	Medium	No change

TABLE A.18
Unique Climate Adaptation Strategies Included in the Analysis: Provide Public Safety

Climate Adaptation Strategy	IPCC Adaptation Category	Citation	Strength of Evidence	Effectiveness	Feasibility	Cyber Vulnerability
Build a volunteer reserve corps	Services	Frasca, 2010 U.S. Fire Administration, 2008 Ready.gov, undated.	Medium	Minor	High	No change
Develop and exercise mutual aid agreements to cope with demand surge during disasters	Services	Harrington and Cole, 2022 National Emergency Management Association, undated	Strong	Major	High	No change
Develop and exercise plans for relocation and repositioning when weather events threaten facilities	Behavioral	U.S. Fire Administration, 2008	Strong	Moderate	High	No change
Increase staffing	Services	Quinton, 2021 American Ambulance Association, 2021 Friese, 2017	Strong	Major	High	No change

TABLE A.19

Unique Climate Adaptation Strategies Included in the Analysis: Supply Water

Climate Adaptation Strategy	IPCC Adaptation Category	Citation	Strength of Evidence	Effectiveness	Feasibility	Cyber Vulnerability
Acquire and manage ecosystems and other land conservation approaches to benefit water utilities and water supply (including landscape nature-based solutions)	Ecosystem-based	EPA, undated c	Strong	Moderate	Medium	No change
Build flood barriers to protect infrastructure or relocate facilities to higher elevations	Engineered and built environment	EPA, undated c	Strong	Moderate	Medium	No change
Build infrastructure needed for aquifer storage and recovery and increased water storage capacity	Engineered and built environment	EPA, undated c	Strong	Moderate	Low	No change
Diversify options for water supply and expand current sources, including facilities to recycle water	Engineered and built environment	EPA, undated c	Strong	Major	Low	No change
Finance and facilitate systems to recycle water	Engineered and built environment	EPA, undated c	Strong	Major	Low	No change
Implement natural and green infrastructure on site and in municipalities (including neighborhood nature-based solutions)	Engineered and built environment; Ecosystem-based	EPA, undated c	Strong	Moderate	Medium	No change
Implement policies and procedures for post-fire repairs	Government policies and programs	EPA, undated c	Strong	Minor	Medium	No change
Implement policies and procedures for post-flood repairs	Government policies and programs	EPA, undated c	Strong	Minor	Medium	No change
Implement saltwater intrusion barriers and aquifer recharge	Engineered and built environment	EPA, undated c	Strong	Moderate	Medium	No change
Implement water conservation programs to reduce water demand	Government policies and programs	EPA, undated c	Strong	Moderate	High	No change
Implement watershed management	Government policies and programs Ecosystem-based	EPA, undated c	Strong	Moderate	Low*	No change

Table A.19—Continued

Climate Adaptation Strategy	IPCC Adaptation Category	Citation	Strength of Evidence	Effectiveness	Feasibility	Cyber Vulnerability
Improve modeling for electricity and agricultural and other irrigation water demands	Technological	EPA, undated c	Strong	Moderate	Medium	No change
Install low-head dam for saltwater wedge and freshwater pool separation	Engineered and built environment	EPA, undated c	Strong	Moderate	Medium	No change
Integrate flood management and modeling into land-use planning	Government policies and programs	EPA, undated c	Strong	Moderate	High	No change
Plan and establish alternative or on-site power supply	Engineered and built environment	EPA, undated c	Strong	Moderate	Medium	Adds
Practice conjunctive use	Ecosystem-based Engineered and built environment	EPA, undated c	Strong	Moderate	Medium	No change
Practice water conservation and demand management by implementing public outreach efforts	Educational	EPA, undated c	Strong	Moderate	High	No change
Update drought contingency plans	Government policies and programs	EPA, undated c	Strong	Minor	High	No change
Update fire models and practice fire management plans	Government policies and programs	EPA, undated c	Strong	Moderate	High	No change

NOTE: When a strategy listed in the tables applied to more than one impact pathway, and the assessments differed, we used the more conservative value and indicated this with an asterisk.

TABLE A.20
Unique Climate Adaptation Strategies Included in the Analysis: Transmit Electricity

Climate Adaptation Strategy	IPCC Adaptation Category	Citation	Strength of Evidence	Effectiveness	Feasibility	Cyber Vulnerability
Develop advanced visualization and information systems	Informational	Panteli and Mancarella, 2015	Strong	Moderate	High	Adds
Employ bushfire reduction and prescribed fire	Ecosystem-based	Noble et al., 2014	Strong	Moderate	Medium	No change
Deploy distributed generation: distributed PV, microgrids, and minigrids	Technological	Gholami, Aminifar, and Shahidehpour, 2016	Strong	Minor*	Medium	Adds
Develop smart grids	Technological	Cox et al., 2017	Strong	Minor*	Medium	Adds
Elevate electrical infrastructure	Engineered and built environment	Linnenluecke, Stathakis, and Griffiths, 2011	Strong	Major	Medium	No change
Implement enhanced demand-side efficiency and reliability	Educational Behavioral Laws and regulations	Cox et al., 2017	Medium	Minor*	Medium*	No change
Implement enhanced supply-side efficiency and reliability	Technological	Stephens et al., 2013	Strong	Minor	Medium	Reduces
Establish and adopt equipment design standards	Laws and regulations	Allen-Dumas, Binita, and Cunliff, 2019	Medium	Major	Medium	No change
Harden facilities	Engineered and built environment	Panteli and Mancarella, 2015	Strong	Moderate	Medium	No change
Implement demand-sensitive electricity tariffs	Economic	Noble et al., 2014	Strong	Major	Medium	No change
Improve climate value at risk assessment for utilities and assets	Informational	Dietz et al., 2016	Strong	Major	Medium	No change
Improve integrated electricity planning approaches	Informational	Cox et al., 2017	Medium	Major	Medium	No change
Employ improved hazard mapping and monitoring technology	Technological	Noble et al., 2014	Medium	Moderate	Medium	Adds
Increase generation capacity reserve margins	Economic	Allen-Dumas, Binita, and Cunliff, 2019	Medium	Moderate	Medium	Adds
Perform integrated coastal zone management	Government policies and programs	Noble et al., 2014	Strong	Moderate	Medium	No change

Table A.20—Continued

Climate Adaptation Strategy	IPCC Adaptation Category	Citation	Strength of Evidence	Effectiveness	Feasibility	Cyber Vulnerability
Maintain backup components and materials	Engineered and built environment	Panteli and Mancarella, 2015	Strong	Minor	High	No change
Enhance resource efficiency	Behavioral	Stephens et al., 2013	Strong	Minor	Medium	Reduces
Leverage national and regional adaptation plans	Government policies and programs	Noble et al., 2014	Strong	Major	Low	No change
Relocate electrical infrastructure	Engineered and built environment	Linnenluecke, Stathakis, and Griffiths, 2011	Strong	Major	Medium	No change
Build sea walls and coastal protection structures	Engineered and built environment	Noble et al., 2014	Strong	Moderate	Low	No change
Strengthen utility mutual aid agreements	Services	Harrington and Cole, 2022	Strong	Moderate	High	No change
Clear transmission line rights-of-way (e.g., tree trimming, vegetation management)	Ecosystem-based	Panteli and Mancarella, 2015	Strong	Minor	High	No change
Install underground distribution conductor	Engineered and built environment	Taylor and Roald, 2021	Strong	Major	Medium	No change
Improve underground distribution conductor code	Laws and regulations	Taylor and Roald, 2021	Strong	Major	Medium	No change
Carry out vulnerability assessments	Informational	Noble et al., 2014	Strong	Minor	High	No change
Conserve and restore wetlands and floodplains	Ecosystem-based	Noble et al., 2014	Strong	Moderate	Low	No change

NOTE: When a strategy listed in the tables applied to more than one impact pathway, and the assessments differed, we used the more conservative value and indicated this with an asterisk.

TABLE A.21

Unique Climate Adaptation Strategies Included in the Analysis: Transport Cargo and Passengers by Air

Climate Adaptation Strategy	IPCC Adaptation Category	Citation	Strength of Evidence	Effectiveness	Feasibility	Cyber Vulnerability
Build sea defenses and/or perimeter dikes	Engineered and built environment	Burbidge, 2018; Yesudian and Dawson, 2021; Stamos, Mitsakis, and Grau, 2015	Medium	Moderate*	Medium	No change
Modify aircraft design	Technological	Coffel, Thompson, and Horton, 2017; Carpenter, 2018	Weak	Moderate	Medium	No change
Change airline operations (schedules, lower weights)	Services	Coffel, Thompson, and Horton, 2017; Carpenter, 2018	Strong	Moderate	High	No change
Create floating airport	Engineered and built environment	Yesudian and Dawson, 2021	Medium	Minor*	Low	No change
Harden or raise infrastructure	Engineered and built environment	Burbidge, 2018; Yesudian and Dawson, 2021	Medium	Moderate*	Medium	No change
Improve weather detection and communication systems	Technological	Doll et al., 2011	Medium	Major	Medium	Adds
Increase airport drainage capacity	Engineered and built environment	Burbidge, 2018; Stamos, Mitsakis, and Grau, 2015	Medium	Moderate	Medium	No change
Institute probabilistic flight planning	Services	Doll et al., 2011	Medium	Moderate	Medium	No change
Lengthen runways	Engineered and built environment	Coffel, Thompson, and Horton, 2017; Carpenter, 2018	Strong	Major	Medium	No change
Relocate airport assets	Engineered and built environment	Burbidge, 2018; Yesudian and Dawson, 2021	Medium	Major	Low	No change
Relocate electrical infrastructure	Engineered and built environment	Burbidge, 2018	Medium	Major	High	No change
Train staff on emergency procedures	Educational	Doll et al., 2011	Medium	Moderate	High	No change
Update airport building codes/master plan	Laws and regulations	Airport Cooperative Research Program, 2015; Stamos, Mitsakis, and Grau, 2015	Weak	Major	High	No change
Update and coordinate emergency response plans	Laws and regulations	Airport Cooperative Research Program, 2015	Medium	Moderate	High	No change

Table A.21—Continued

Climate Adaptation Strategy	IPCC Adaptation Category	Citation	Strength of Evidence	Effectiveness	Feasibility	Cyber Vulnerability
Use natural barriers	Ecosystem-based	Burbidge, 2018	Medium	Moderate	Medium	No change
Use tarmac mix that accelerates drainage	Engineered and built environment	Stamos, Mitsakis, and Grau, 2015	Medium	Major	High	No change

NOTE: When a strategy listed in the tables applied to more than one impact pathway, and the assessments differed, we used the more conservative value and indicated this with an asterisk.

TABLE A.22
Unique Climate Adaptation Strategies Included in the Analysis: Transport Cargo and Passengers by Rail

Climate Adaptation Strategy	IPCC Adaptation Category	Citation	Strength of Evidence	Effectiveness	Feasibility	Cyber Vulnerability
Build flood barriers to protect tracks	Engineered and built environment	Stamos, Mitsakis, and Grau, 2015	Medium	Major	Medium	No change
Conduct preventive maintenance on track	Engineered and built environment	Doll, Klug, and Enei, 2014	Medium	Moderate	Medium	No change
Develop flood warning plans for bridges and culverts	Informational	Marteaux, 2016	Medium	Moderate	High	No change
Elevate track and stations at risk	Engineered and built environment	Stamos, Mitsakis, and Grau, 2015	Medium	Major	Medium	No change
Enhance slope stability	Engineered and built environment	Palin et al., 2021; Doll et al., 2011	Medium	Major	Medium	No change
Improve pumping and ventilation in tunnels	Engineered and built environment	Stamos, Mitsakis, and Grau, 2015	Medium	Major	Medium	No change
Improve real-time monitoring of infrastructure	Technological	Stamos, Mitsakis, and Grau, 2015	Medium	Minor	High	Adds
Improve weather detection and communication systems	Technological	Doll, Klug, and Enei, 2014	Medium	Moderate	Medium	Adds
Include climate impacts in long-range planning	Government policies and programs	Doll et al., 2011	Medium	Moderate	High	No change
Maintain retaining walls	Engineered and built environment	Palin et al., 2021	Medium	Major	Medium	No change
Maintain vegetation near track to reduce possible storm damage	Ecosystem-based	Palin et al., 2021	Medium	Moderate	High	No change
Move tracks and stations at risk	Engineered and built environment	Mulkern, 2022	Weak	Major	Low	No change
Shore up underwater bridge supports	Engineered and built environment	Palin et al., 2021	Medium	Major	Medium	No change

Table A.22—Continued

Climate Adaptation Strategy	IPCC Adaptation Category	Citation	Strength of Evidence	Effectiveness	Feasibility	Cyber Vulnerability
Strengthen infrastructure codes and standards	Engineered and built environment	Palin et al., 2021	Medium	Major	Medium	No change
Upgrade drainage systems	Engineered and built environment	Palin et al., 2021 Stamos, Mitsakis, and Grau, 2015	Medium	Major	Medium	No change
Use pile construction for buildings with electrical equipment	Engineered and built environment	Doll et al., 2011	Medium	Major	Medium	No change

TABLE A.23
Unique Climate Adaptation Strategies Included in the Analysis: Transport Cargo and Passengers by Road

Climate Adaptation Strategy	IPCC Adaptation Category	Citation	Strength of Evidence	Effectiveness	Feasibility	Cyber Vulnerability
Build flood barriers to protect roads	Engineered and built environment	Doll et al., 2011; Stamos, Mitsakis, and Grau, 2015	Medium	Major	Medium	No change
Design and maintain drainage systems	Engineered and built environment	Doll et al., 2011	Medium	Major	High	No change
Elevate roads at risk	Engineered and built environment	Stamos, Mitsakis, and Grau, 2015	Medium	Moderate	Medium	No change
Improve pumping and ventilation in tunnels	Engineered and built environment	Stamos, Mitsakis, and Grau, 2015	Medium	Moderate	High	No change
Improve slope stability	Engineered and built environment	Doll et al., 2011	Medium	Moderate	High	No change
Improve vehicle communications and technology	Technological	Doll et al., 2011	Medium	Moderate	High	Adds
Maintain pavements and conduct preventive maintenance	Engineered and built environment	Doll et al., 2011	Medium	Moderate	High	No change
Maintain vegetation to reduce risks of debris and erosion	Ecosystem-based	Doll et al., 2011	Medium	Major	High	No change
Move roads at risk	Engineered and built environment	Doll et al., 2011	Medium	Major	Medium	No change
Plan for climate-sensitive road alignments	Government policies and programs	Doll et al., 2011	Medium	Major	Medium	No change
Provide real-time information to drivers	Informational	Stamos, Mitsakis, and Grau, 2015	Medium	Minor	High	Adds
Update and coordinate emergency response plans	Laws and regulations	Stamos, Mitsakis, and Grau, 2015	Medium	Moderate	High	No change
Strengthen infrastructure codes and standards for road design, including bridges and tunnels	Laws and regulations	Doll et al., 2011	Medium	Moderate	High	No change
Use asphalt that improves drainage	Engineered and built environment	Stamos, Mitsakis, and Grau, 2015	Medium	Moderate	High	No change
Use more-resilient pavement materials	Engineered and built environment	Doll et al., 2011	Medium	Moderate	High	No change

TABLE A.24
Unique Climate Adaptation Strategies Included in the Analysis: Transport Cargo and Passengers by Vessel

Climate Adaptation Strategy	IPCC Adaptation Category	Citation	Strength of Evidence	Effectiveness	Feasibility	Cyber Vulnerability
Build and maintain spur dikes	Engineered and built environment	Doll et al., 2011	Medium	Moderate	Low	No change
Build storm surge barriers	Engineered and built environment	Doll et al., 2011 Stamos, Mitsakis, and Grau, 2015	Medium	Moderate	Low	No change
Conduct preventive maintenance	Engineered and built environment	Doll et al., 2011 Stamos, Mitsakis, and Grau, 2015	Medium	Moderate	High	No change
Design more-resilient ships	Technological	Doll et al., 2011 Stamos, Mitsakis, and Grau, 2015	Medium	Major	High	No change
Elevate harbor infrastructure	Engineered and built environment	Stamos, Mitsakis, and Grau, 2015	Medium	Major	Medium	No change
Improve infrastructure monitoring	Technological	Stamos, Mitsakis, and Grau, 2015	Medium	Minor	Medium	Adds
Improve river condition monitoring	Technological	Stamos, Mitsakis, and Grau, 2015	Medium	Moderate	Medium	Adds
Improve transshipment infrastructure	Engineered and built environment	Stamos, Mitsakis, and Grau, 2015	Medium	Minor	Medium	No change
Install flood gates	Engineered and built environment	Doll et al., 2011	Medium	Major	Low	No change
Install water breaks	Engineered and built environment	Stamos, Mitsakis, and Grau, 2015	Medium	Major	Medium	No change
Prepare alternative modes in case of disruption	Services	Doll et al., 2011	Medium	Minor*	High	No change
Provide adequate fendering systems for lighter-weight vessels	Engineered and built environment	Stamos, Mitsakis, and Grau, 2015	Medium	Minor	Medium	No change
Reinforce port infrastructure (e.g., docks and cranes)	Engineered and built environment	Stamos, Mitsakis, and Grau, 2015	Medium	Major	Medium	No change
Relocate harbor infrastructure	Engineered and built environment	Stamos, Mitsakis, and Grau, 2015	Medium	Major	Medium	No change
Remove sediment from channels	Engineered and built environment	Stamos, Mitsakis, and Grau, 2015	Medium	Major	Medium	No change

Table A.24—Continued

Climate Adaptation Strategy	IPCC Adaptation Category	Citation	Strength of Evidence	Effectiveness	Feasibility	Cyber Vulnerability
Update and coordinate emergency response plans	Laws and regulations	Doll et al., 2011	Medium	Moderate	High	No change

NOTE: When a strategy listed in the tables applied to more than one impact pathway, and the assessments differed, we used the more conservative value and indicated this with an asterisk.

TABLE A.25
Unique Climate Adaptation Strategies Included in the Analysis: Transport Passengers by Mass Transit

Climate Adaptation Strategy	IPCC Adaptation Category	Citation	Strength of Evidence	Effectiveness	Feasibility	Cyber Vulnerability
Design alignment to optimize energy use	Engineered and built environment	American Public Transportation Association, 2011	Weak	Minor	Low	No change
Improve energy efficiency	Technological	Tian et al., 2019 Doll et al., 2011	Medium	Moderate	High	No change
Provide incentives to use adaptation measures	Economic	Doll et al., 2011	Weak	Moderate	Medium	No change
Switch to electricity sources that require less water for generation	Technological	American Public Transportation Association, 2011	Weak	Major	Medium	No change

Planning and Decision Tools Identified

Table B.1 lists planning and decision tools that were identified by HSOAC SMEs identified as having relevance to at least one NCF; were developed by the federal government, a university, or other authoritative source; and are available free of charge. While this is not an exhaustive list of all the potentially relevant tools, there are over 50 guides and tools listed that could be used by owner-operators to begin to assess alternative adaptation strategies. Links are available in the list of references.

TABLE B.1

Planning and Decision Tools Identified

Tool/Aid	Developer	Citations
Adaptation and Resilience Planning for Providers of Public Transportation	Southern California Association of Governments	Southern California Association of Governments, undated
Adaptation Tool Kit: Sea-Level Rise and Coastal Land Use	Georgetown Climate Center	Grannis, 2011
Adaptation Workbook	Northern Institute of Applied Climate Science	Northern Institute of Applied Climate Science, undated
Adapting to Urban Heat: A Tool Kit for Local Governments	Georgetown Climate Center	Hoverter, 2012
AgBiz Logic	Oregon State University	Oregon State University, 2022
Airport Climate Risk Operational Screening (ACROS) tool	Transportation Research Board, Airport Cooperative Research Program	Airport Cooperative Research Program, 2015
Assessing Criticality in Transportation Adaptation Planning	U.S. Department of Transportation	Federal Highway Administration, 2014
Build Wildfire Resilience at Water and Wastewater Utilities	EPA	EPA, undated b
Building America Solution Center—Disaster Resistance Tool	U.S. Department of Energy	U.S. Department of Energy, undated
Building Code Adoption Tracking Portal	FEMA	FEMA, undated
Climate Change Handbook for Regional Water Planning	EPA	EPA, undated d
Climate Resilience Evaluation and Awareness Tool (CREAT)	EPA	EPA, 2022a
Coastal Resilience Evaluation and Siting Tool (CREST)	National Fish and Wildlife Foundation	National Fish and Wildlife Foundation, undated
Community Resilience Portal	HUD	O'Grady et al., 2022
Community-Based Climate Adaptation Toolkit	Reef Resilience Network	Reef Resilience Network, undated

Table B.1—Continued

Tool/Aid	Developer	Citations
Compendium of Adaptation Approaches	USDA	USDA, undated a
Creating Resilient Water Utilities	EPA	EPA, undated e
Design Guide for Improving Hospital Safety in Earthquakes, Floods, and High Winds (FEMA 577)	FEMA	Arnold et al., 2007
Drought Response and Recovery Guide for Water Utilities	EPA	EPA, 2018
Enhancing Sustainable Communities with Green Infrastructure Guide	EPA	Kramer, 2014
Environmental Justice Screening and Mapping Tool	EPA	EPA, undated f
Environmental Resilience Institute Toolkit (ERIT)	Environmental Resilience Institute	Environmental Resilience Institute, undated
EPA Flood Resilience Checklist	EPA	EPA, 2014b
EPA Flood Resilience: A Basic Guide for Water and Wastewater Utilities	EPA	EPA, 2014c
EPA National Stormwater Calculator	EPA	EPA, 2022b
Federal Highway Administration Scenario Planning Guidebook	U.S. Department of Transportation Federal Highway Administration	Twaddell et al., 2016
FEMA National Flood Insurance Program Community Rating System	FEMA	FEMA, 2018
Incorporating the Costs and Benefits of Adaptation Measures in Preparation for Extreme Weather Events and Climate Change—Guidebook	National Cooperative Highway Research Program	National Cooperative Highway Research Program, 2020
International Climate Change Adaptation Framework for Road Infrastructure	World Road Association (PIARC)	Toplis et al., 2015
Maintenance Decision Support System (MDSS)	Federal Highway Administration	Federal Highway Administration, undated
Managed Retreat Toolkit	Georgetown Climate Center	Georgetown Climate Center, undated
National Fish, Wildlife, and Plants Climate Adaptation Strategy	Association of Fish and Wildlife Agencies	National Fish, Wildlife, and Plants Climate Adaptation Network, 2021
Natural Infrastructure Opportunities Tool	U.S. Army Corps of Engineers	U.S. Army Corps of Engineers, undated
Primary Protection: Enhancing Health Care Resilience for a Changing Climate	HHS	Guenther and Balbus, 2014
Rapid Vulnerability & Adaptation Tool for Climate-Informed Community Planning	EcoAdapt	EcoAdapt, 2021
Regional Adaptation Collaborative Toolkit	Alliance of Regional Collaboratives for Climate Adaptation (California based)	Alliance of Regional Collaboratives for Climate Adaptation, undated
Resilient Strategies Guide for Water Utilities	EPA	EPA, 2022c

Table B.1—Continued

Tool/Aid	Developer	Citations
Rolling Easements Primer	EPA	Titus, 2011
Scenario Planning for Crops and Cattle	Institute of Agriculture and Natural Resources	Institute of Agriculture and Natural Resources, undated
Smart Growth Fixes for Climate Adaptation and Resilience	EPA	EPA, 2017b
Storm Water Management Model (SWMM)	EPA	EPA, undated g
Synthesis of Adaptation Options for Coastal Areas	EPA	EPA, 2009
Tribal Climate Change Adaptation Planning Toolkit	Institute for Tribal Environmental Professionals	Institute for Tribal Environmental Professionals, undated
Urban Adaptation Assessment	Notre Dame Global Adaptation Initiative	Notre Dame Global Adaptation Initiative, undated
USGCRP Reports & Resources website	USGCRP	USGCRP, undated
Vulnerability Assessment and Adaptation Framework	Federal Highway Administration	Filosa et al., 2017
Vulnerability, Consequences, and Adaptation Planning Scenarios	Social and Environmental Research Institute, Inc.	Social and Environmental Research Institute, Inc., undated

Subject-Matter Expertise

Table C.1 lists the lead analyst(s), or SMEs, and reviewers with subject-matter expertise for each NCF. The respective roles of these positions are described in Chapter Two. Quentin Hodgson also provided expertise on cybersecurity.

TABLE C.1

Lead Analysts and Subject-Matter Experts

Sector	NCF	SME Analysts	SME Reviewers
Agriculture	Produce and Provide Agricultural Products and Services	Patricia Stapleton	Karen Sudkamp
Energy	Distribute Electricity	Liam Regan	Ismael Arciniegas Rueda Kelly Klima
	Exploration and Extraction of Fuels	Liam Regan	Ismael Arciniegas Rueda Kelly Klima
	Generate Electricity	Liam Regan	Ismael Arciniegas Rueda Kelly Klima
	Transmit Electricity	Liam Regan	Ismael Arciniegas Rueda Kelly Klima
Government and Social Services	Educate and Train	Timothy Gulden	Shelly Culbertson
	Enforce Law	Dic Donohue	John Hollywood Edward Chan
	Prepare for and Manage Emergencies	Edward Chan	Jason Barnosky
	Provide Housing	Michael Wilson	Gary Cecchine Matt Baird
	Provide Medical Care	Edward Chan	Mahshid Abir Shira Fischer
	Provide Public Safety	Edward Chan	John Hollywood
Industry	Maintain Supply Chains	Tobias Sytsma	Ellen Pint
	Manufacture Equipment	Tobias Sytsma	Ellen Pint
	Produce Chemicals	Tobias Sytsma	Ellen Pint
	Provide Insurance Services	Timothy Gulden	R. J. Briggs
Infrastructure	Develop and Maintain Public Works and Services	Rahim Ali	Ellen Pint
	Provide and Maintain Infrastructure	Tobias Sytsma Rahim Ali	Ellen Pint

Table C.1—Continued

Sector	NCF	SME Analysts	SME Reviewers
Transportation	Transport Cargo and Passengers by Air	Liisa Ecola	Nahom Beyene Jeremy Eckhause
	Transport Cargo and Passengers by Rail	Liisa Ecola	Nahom Beyene Jeremy Eckhause
	Transport Cargo and Passengers by Road	Liisa Ecola	Nahom Beyene Jeremy Eckhause
	Transport Cargo and Passengers by Vessel	Liisa Ecola	Nahom Beyene Jeremy Eckhause
	Transport Passengers by Mass Transit	Liisa Ecola	Nahom Beyene Jeremy Eckhause
Water and Waste Management	Manage Wastewater	Chelsea Kolb	Beth Lachman Susan Resetar
	Supply Water	Chelsea Kolb	Beth Lachman Susan Resetar
	Manage Hazardous Materials	Michael Wilson	Gary Cecchine Susan Resetar

Abbreviations

CISA	Cybersecurity and Infrastructure Security Agency
EPA	U.S. Environmental Protection Agency
FEMA	Federal Emergency Management Agency
HHS	U.S. Department of Health and Human Services
HSOAC	Homeland Security Operational Analysis Center
HUD	U.S. Department of Housing and Urban Development
IPCC	Intergovernmental Panel on Climate Change
NCF	National Critical Function
NOAA	National Oceanographic and Atmospheric Administration
OECD	Organisation for Economic Co-operation and Development
OSHA	Occupational Safety and Health Administration
PV	photovoltaic cells
SLTT	state, local, tribal, and territorial governments
SME	subject-matter expert
UNEP	United Nations Environment Programme
USDA	U.S. Department of Agriculture
USGCRP	U.S. Global Change Research Program

References

Abbott, Chris, *An Uncertain Future: Law Enforcement, National Security and Climate Change*, Oxford Research Group, 2008.

Acosta, Joie D., Anita Chandra, Jaime Madrigano, "An Agenda to Advance Integrative Resilience Research and Practice: Key Themes from a Resilience Roundtable," *RAND Health Quarterly*, Vol. 7., No. 1, January 1, 2017.

Adaptation Clearinghouse, "Welcome to the Adaptation Clearinghouse," undated. As of August 9, 2022: https://www.adaptationclearinghouse.org/

Adelaine, Sabrina A., Mizuki Sato, Yufang Jin, and Hilary Godwin, "An Assessment of Climate Change Impacts on Los Angeles (California USA) Hospitals, Wildfires Highest Priority." *Prehospital Disaster Medicine*, Vol. 32, No. 5, 2017.

Advisen, "Chemical Companies: Minimizing the Risks of Supply Chain Disruptions," white paper, January 2013.

Agrawal, K. C., *Electrical Power Engineering: Reference and Applications Handbook*, Mumbai City, Maharashtra: One Point Six Technologies, 2020.

Airport Cooperative Research Program, *Climate Change Adaptation Planning: Risk Assessment for Airports*, National Academies Press, 2015. As of September 7, 2022: https://nap.nationalacademies.org/catalog/23461/ climate-change-adaptation-planning-risk-assessment-for-airports

Alexandratos, Spiro D., Naty Barak, Diana Bauer, F. Todd Davidson, Brian R. Gibney, Susan S. Hubbard, Hessy L. Taft, and Paul Westerhof, "Sustaining Water Resources: Environmental and Economic Impact," *ACS Sustainable Chemistry & Engineering*, Vol. 7, No. 3, February 4, 2019.

Allen-Dumas, Melissa, K. C. Binita, and Colin I. Cunliff, *Extreme Weather and Climate Vulnerabilities of the Electric Grid: A Summary of Environmental Sensitivity Quantification Methods*, Oak Ridge National Laboratory, 2019.

Alliance of Regional Collaboratives for Climate Adaptation, "Regional Adaptation Collaborative Toolkit," webpage, undated. As of August 15, 2022: https://arccacalifornia.org/toolkit/

Álvarez-Berríos, Nora L., Sarah L. Wiener, Kathleen A. McGinley, Angela B. Lindsey, and William A. Gould, "Hurricane Effects, Mitigation, and Preparedness in the Caribbean: Perspectives on High Importance–Low Prevalence Practices from Agricultural Advisors," *Journal of Emergency Management*, Vol. 19, No. 8, 2021.

American Ambulance Association, "Congressional Letter on the EMS Workforce Shortage," October 1, 2021. As of August 9, 2022: https://ambulance.org/2021/10/04/workforceshortage/

American Farm Bureau Federation, "Hurricane Ida: Direct Agricultural Impacts and Larger Implications of Flooding," webpage, September 15, 2021.

American Flood Coalition, "Flood Insurance and Risk Rating 2.0: Everything You Need to Know in Five Minutes," webpage, September 7, 2021.

American Hospital Association, "Strengthening the Health Care Workforce," fact sheet, November 2021.

American Public Transportation Association, "Transit Sustainability Guidelines: Framework for Approaching Sustainability and Overview of Best Practices," APTA SUDS-CC-RP-004-11, March 31, 2011.

American Water Works Association, *Emergency Planning for Water and Wastewater Utilities*, 5th ed., 2018.

American Water Works Association, *Drought Preparedness and Response*, 2nd ed., 2019.

American Water Works Association, *Climate Action Plans—Adaptive Management Strategies for Utilities*, 2021.

Arent, Douglas J., Richard S. J. Tol, Eberhard Faust, Joseph P. Hella, Surender Kumar, Kenneth M. Strzepek, Ferenc L. Tóth, and Denghua Yan, "Key Economic Sectors and Services," Ch. 10 in C. B. Field, V. R. Barros, D. J. Dokken, K. J. Mach, M. D. Mastrandrea, T. E. Bilir, M. Chatterjee, K. L. Ebi, Y. O. Estrada, R. C. Genova, B. Girma, E. S. Kissel, A. N. Levy, S. MacCracken, P. R. Mastrandrea, and L. L. White, eds., *Climate Change 2014: Impacts, Adaptation, and Vulnerability,* Part A: *Global and Sectoral Aspects,* Contribution of Working Group II to the Fifth Assessment Report of the Intergovernmental Panel on Climate Change, Cambridge University Press, 2014. As of August 9, 2022:
https://www.ipcc.ch/site/assets/uploads/2018/02/WGIIAR5-Chap10_FINAL.pdf

Arnold, Christopher, William B. Holmes, Rebecca C. Quinn, Thomas L. Smith, Bogdan Srdanovic, and Wilbur Tussler, *Design Guide for Improving Hospital Safety in Earthquakes, Floods, and High Winds,* Federal Emergency Management Agency, June 2007. As of September 7, 2022:
https://www.wbdg.org/ffc/dhs/criteria/fema-577

Baker, Earl J., Robert E. Deyle, Timothy S. Chapin, and John B. Richardson, "Are We Any Safer? Comprehensive Plan Impacts on Hurricane Evacuation and Shelter Demand in Florida," *Coastal Management,* Vol. 36, No. 3, 2008.

Baskin, Kara, "Supply Chain Resilience in the Era of Climate Change," webpage, MIT Sloan School of Management, February 11, 2020.

Baxter, Julie, Karen Helbrecht, Stacy Franklin Robinson, Sara Reynolds, Adam Reeder, and Hilary Kendro, *Mitigation Ideas: A Resource for Reducing Risk to Natural Hazards,* Federal Emergency Management Agency, January 2013.

Bravo, Raissa Zurli Bittencourt, Adriana Leiras, and Fernando Luiz Cyrino Oliveira, "Mitigation and Prevention of Droughts: A Systematic Literature Review," in Adriana Leiras, Carlos Alberto González-Calderón, Irineu de Brito Junior, Sebastián Villa, and Hugo Tsugunobu Yoshida Yoshizaki, eds., *Operations Management for Social Good: 2018 POMS International Conference in Rio,* Springer, 2020.

Bressers, Hans, Nanny Bressers, and Corinne Larrue, *Governance for Drought Resilience: Land and Water Drought Management in Europe,* Springer, 2016.

Brody, Sarah, Matt Rogers, and Giulia Siccardo, "Why, and How, Utilities Should Start to Manage Climate-Change Risk," McKinsey and Company, April 2019.

Brugmann, Jeb, *Financing the Resilient City: A Demand Driven Approach to Development, Disaster Risk Reduction and Climate Adaptation,* ICLEI—Local Governments for Sustainability, 2011.

Bulkeley, Harriet, and Rafael Tuts, "Understanding Urban Vulnerability, Adaptation and Resilience in the Context of Climate Change," *Local Environment,* Vol. 18, No. 6, 2013.

Burbidge, Rachel, "Adapting Aviation to a Changing Climate: Key Priorities for Action," *Journal of Air Transport Management,* Vol. 71, August 2018.

Busby, Joshua W., Kyri Baker, Morgan D. Bazilian, Alex Q. Gilbert, Emily Grubert, Varun Rai, Joshua D. Rhodes, Sarang Shidore, Caitlin A. Smith, and Michael E. Webber, "Cascading Risks: Understanding the 2021 Winter Blackout in Texas," *Energy Research and Social Science,* Vol. 77, July 2021.

Busch, Timo, "Organizational Adaptation to Disruptions in the Natural Environment: The Case of Climate Change," *Scandinavian Journal of Management,* Vol. 27, No. 4, 2011.

Cai, Ximing, Ruijie Zeng, Won Hee Kang, and Junho Song, "Strategic Planning for Drought Mitigation Under Climate Change," *Journal of Water Resources Planning and Management,* Vol. 141, No. 9, September 2015.

California Climate and Agriculture Network, "Wildfire Mitigation Strategies," webpage, undated. As of August 10, 2022:
https://calclimateag.org/solutions/wildfire-mitigation-strategies/

Calma, Justine, "Texas' Natural Gas Production Just Froze Under Pressure," webpage, The Verge, February 17, 2021.

Carpenter, Brandon T., *An Overview and Analysis of the Impacts of Extreme Heat on the Aviation Industry,* University of Tennessee at Knoxville, Chancellor's Honors Program Projects, 2018.
https://trace.tennessee.edu/cgi/viewcontent.cgi?article=3188&context=utk_chanhonoproj

Chandra, Anita, Joie Acosta, Stefanie Stern, Lori Uscher-Pines, Malcolm V. Williams, Douglas Yeung, Jeffrey Garnett, and Lisa S. Meredith, *Building Community Resilience to Disasters: A Way Forward to Enhance National Health Security*, Santa Monica, Calif.: RAND Corporation, 2011. As of August 10, 2022: https://www.rand.org/pubs/technical_reports/TR915.html

Charnley, Susan, Melissa R. Poe, Alan A. Ager, Thomas A. Spies, Emily K. Platt, and Keith A. Olsen, "A Burning Problem: Social Dynamics of Disaster Risk Reduction Through Wildfire Mitigation," *Human Organization*, Vol. 74, No. 4, 2015.

CISA—*See* Cybersecurity and Infrastructure Security Agency.

Coffel, Ethan D., Terence R. Thompson, and Radley M. Horton, "The Impacts of Rising Temperatures on Aircraft Takeoff Performance," *Climatic Change*, Vol. 144, No. 2, 2017.

Costa, Sergio A., Ilias Kavouras, Nevin Cohen, and Terry T. K. Huang, "Moving Education Online During the COVID-19 Pandemic: Thinking Back and Looking Ahead," *Frontiers in Public Health*, Vol. 9, October 25, 2021.

Cox, Sadie, Eliza Hotchkiss, Dan Bilello, Andrea Watson, Alison Holm, and Jennifer Leisch, *Bridging Climate Change Resilience and Mitigation in the Electricity Sector Through Renewable Energy and Energy Efficiency: Emerging Climate Change and Development Topics for Energy Sector Transformation*, U.S. Agency for International Development and National Renewable Energy Laboratory, November 2017.

Culp, Kenneth, Shalome Tonelli, Sandra L. Ramey, Kelley Donham, and Laurence Fuortes, "Preventing Heat-Related Illness Among Hispanic Farmworkers," *AAOHN Journal*, Vol. 59, No. 1, 2011.

Cutter, Susan, Balgis Osman-Elasha, John Campbell, So-Min Cheong, Sabrina McCormick, Roger Pulwarty, Seree Supratid, and Gina Ziervogel, "Managing the Risks from Climate Extremes at the Local Level," in Christopher B. Field, Vicente Barros, Thomas F. Stocker, Qin Dahe, David Jon Dokken, Gian-Kasper Plattnern, Kristie L. Ebin, Simon K. Allenn, Michael D. Mastrandrean, Melinda Tignorn, Katharine J. Mach, and Pauline M. Midgley, eds., *Managing the Risks of Extreme Events and Disasters to Advance Climate Change Adaptation, Special Report of Working Groups I and II of the Intergovernmental Panel on Climate Change (IPCC)*, Cambridge University Press, 2012.

Cybersecurity and Infrastructure Security Agency, "National Critical Functions," fact sheet, undated.

Cybersecurity and Infrastructure Security Agency, "Hurricane-Related Scams," webpage, June 2020. As of August 11, 2022: https://www.cisa.gov/uscert/ncas/current-activity/2020/06/01/hurricane-related-scams

Deryugina, Tatyana, and Megan Konar, "Impacts of Crop Insurance on Water Withdrawals for Irrigation," *Advances in Water Resources*, Vol. 110, December 2017.

Dietz, Simon, Alex Bowen, Charlie Dixon, and Philip Gradwell, "'Climate Value at Risk' of Global Financial Assets," *Nature Climate Change*, Vol. 6, No. 7, July 2016.

Dilling, Lisa, Meaghan E. Daly, William R. Travis, Olga V. Wilhelmi, and Roberta A. Klein, "The Dynamics of Vulnerability: Why Adapting to Climate Variability Will Not Always Prepare Us for Climate Change," *Wiley Interdisciplinary Reviews: Climate Change*, Vol. 6, No. 4, July–August 2015.

Dodgen, Daniel, Darrin Donato, Nancy Kelly, Annette La Greca, Joshua Morganstein, Joseph Reser, Josef Ruzek, Shulamit Schweitzer, Mark M. Shimamoto, Kimberly Thigpen Tart, and Robert Ursano, "Mental Health and Well-Being," in Allison Crimmins et al., eds. *The Impacts of Climate Change on Human Health in the United States: A Scientific Assessment*, U.S. Global Change Research Program, 2016.

Doherty, R. E., and H. H. Dewey, "Fundamental Considerations of Power Limits of Transmission Systems," *Journal of the A.I.E.E.*, Vol. 44, No. 10, October 1925.

Doll, Claus, Stefan Klug, Ina Partzsch, Riccardo Enei, Verena Pelikan, Norbert Sedlacek, Hedi Maurer, Loreta Rudzikaite, Anestis Papanikolaou, and Vangelis Mitsakis, *Adaptation Strategies in the Transport Sector,* Weather Extremes: Assessment of the Impacts on Transport Systems and Hazards for European Regions, 2011.

Doll, Claus, Stefan Klug, and Riccardo Enei, "Large and Small Numbers: Options for Quantifying the Costs of Extremes on Transport Now and in 40 Years," *Natural Hazards*, Vol. 72, No. 1, 2014.

Donovan, Geoffrey H., and Jeffrey P. Prestemon, "The Effect of Trees on Crime in Portland, Oregon," *Environment and Behavior*, Vol. 44, No. 1, 2012.

EcoAdapt, "Rapid Vulnerability & Adaptation Tool for Climate-Informed Community Planning," brochure, Spring 2021. As of September 7, 2022:
http://ecoadapt.org/data/documents/EcoAdapt_RVAT_FillableWorksheets.pdf

Environmental Resilience Institute, "Environmental Resilience Institute Toolkit," Indiana University, undated. As of September 7, 2022:
https://eri.iu.edu/erit/

EPA—*See* U.S. Environmental Protection Agency.

Etemadi, Niloofar, Yari Borbon-Galvez, Fernanda Strozzi, and Tahereh Etemadi, "Supply Chain Disruption Risk Management with Blockchain: A Dynamic Literature Review," *Information*, Vol. 12, No. 2, 2021.

Executive Order 14008, *Tackling the Climate Crisis at Home and Abroad*, January 27, 2021.

Extreme Weather Response Task Force, "The New Normal: Combating Storm-Related Extreme Weather in New York City," Office of the Deputy Mayor for Administration, New York City, 2021. As of August 15, 2022:
https://www1.nyc.gov/assets/orr/pdf/publications/WeatherReport.pdf

Federal Emergency Management Agency, "Nationwide Building Code Adoption Tracking," webpage, undated. As of August 15, 2022:
https://www.fema.gov/emergency-managers/risk-management/building-science/bcat

Federal Emergency Management Agency, *Design Guide for Improving Hospital Safety in Earthquakes, Floods, and High Winds*, June 2007.

Federal Emergency Management Agency, *Safer, Stronger, Smarter: A Guide to Improving School Natural Hazard Safety*, June 2017.

Federal Emergency Management Agency, "Community Rating System: A Local Official's Guide to Saving Lives, Preventing Property Damage, and Reducing the Cost of Flood Insurance," brochure, 2018. As of September 7, 2022:
https://www.fema.gov/floodplain-management/community-rating-system

Federal Emergency Management Agency, *Building Codes Saves: A Nationwide Study*, November 2020.

Federal Emergency Management Agency, *Region 10 Guide to Expanding Mitigation: Connecting with Agriculture and Food Systems*, May 2021.

Federal Highway Administration, "Maintenance Decision Support (MDSS) Showcase," webpage, U.S. Department of Transportation, undated. As of August 15, 2022:
https://ops.fhwa.dot.gov/weather/seminars/mdss_showcase/index.htm

Federal Highway Administration, "Assessing Criticality in Transportation Adaptation Planning," webpage, May 25, 2014. As of September 7, 2022:
https://www.fhwa.dot.gov/environment/sustainability/resilience/tools/criticality_guidance/index.cfm

FEMA—*See* Federal Emergency Management Agency.

Filosa, Gina, Amy Plovnick, Leslie Stahl, Rawlings Miller, and Don Pickrell, *Vulnerability Assessment and Adaptation Framework*, 3rd ed., U.S. Department of Transportation, Federal Highway Administration, December 2017. As of September 7, 2022:
https://www.fhwa.dot.gov/environment/sustainability/resilience/adaptation_framework/index.cfm

Frank, Thomas, "Why the U.S. Disaster Agency Is Not Ready for Catastrophes," *Scientific American*, August 20, 2019.

Frasca, Dominic R., "The Medical Reserve Corps as Part of the Federal Medical and Public Health Response in Disaster Settings," *Biosecurity and Bioterrorism: Biodefense Strategy, Practice, and Science*, September 2010.

Friese, Greg, "It's Time to Talk Climate Change and Its Effect on First Responders," EMS1, October 10, 2017.

Fu, Xinyu, Zhenghong Tang, Jianjun Wu, and Kevin McMillan, "Drought Planning Research in the United States: An Overview and Outlook," *International Journal of Disaster Risk Science*, Vol. 4, 2013.

Gamble, Janet L., Martha Berger, Karen Bouye, Vince Campbell, Karletta Chief, Kathryn Colon, Allison Crimmins, Barry Flanagan, Cristina Gonzalez-Maddux, et al., "Populations of Concern," in Allison Crimmins et al., eds. *The Impacts of Climate Change on Human Health in the United States: A Scientific Assessment*, U.S. Global Change Research Program, 2016.

Georgetown Climate Center, "Managed Retreat Toolkit," webpage, undated. As of September 7, 2022: https://www.georgetownclimate.org/adaptation/toolkits/managed-retreat-toolkit/introduction.html

Gholami, Amin, Farrokh Aminifar, and Mohammad Shahidehpour, "Front Lines Against the Darkness: Enhancing the Resilience of the Electricity Grid Through Microgrid Facilities," *IEEE Electrification Magazine*, Vol. 4, No. 1, 2016.

Gibbs, Linda, and Cas Holloway, *Hurricane Sandy After Action Report and Recommendations to Mayor Michael R. Bloomberg*, City of New York, May 2013.

Giller, Ken E., Jens A. Andersson, Marc Corbeels, John Kirkegaard, David Mortensen, Olaf Erenstein, and Bernard Vanlauwe, "Beyond Conservation Agriculture," *Frontiers in Plant Science*, Vol. 6, 2015.

Goentzel, Jarrod, and Michael Windle, eds., *Supply Chain Resilience: Restoring Business Operations After a Hurricane: Summary Report*, MIT Center for Transportation & Logistics, 2017.

Goin, Dana E., Kara E. Rudolph, and Jennifer Ahern, "Impact of Drought on Crime in California: A Synthetic Control Approach," *PLoS One*, Vol. 12, No. 10, October 4, 2017.

Gopalakrishnan, Tharani, Md Kamrul Hasan, A. T. M. Sanaul Haque, Sadeeka Layomi Jayasinghe, and Lalit Kumar, "Sustainability of Coastal Agriculture Under Climate Change," *Sustainability*, Vol. 11, No. 24, 2019.

Grannis, Jessica, *Adaptation Tool Kit: Sea-Level Rise and Coastal Land Use*, Georgetown Climate Center, October 2011. As of September 7, 2022: https://www.georgetownclimate.org/files/report/Adaptation_Tool_Kit_SLR.pdf

Grice, Lori, "FBI New Orleans Warns About Hurricane-Related Fraud," press release, Federal Bureau of Investigation New Orleans, August 30, 2021. As of August 11, 2022: https://www.fbi.gov/contact-us/field-offices/neworleans/news/press-releases/fbi-new-orleans-warns-about-hurricane-related-fraud-1

Groves, David G., Edmundo Molina-Perez, Evan Bloom, and Jordan R. Fischbach, "Robust Decision Making (RDM): Application to Water Planning and Climate Policy," in Vincent A. W. J. Marchau, Warren E. Walker, Pieter J. T. M. Bloemen, and Steven W. Popper, eds., *Decision Making Under Deep Uncertainty*, Springer, 2019.

Guenther, Robin, and John Balbus, *Primary Protection: Enhancing Health Care Resilience for a Changing Climate*, U.S. Department of Health and Human Services, December 2014. As of September 7, 2022: https://toolkit.climate.gov/sites/default/files/SCRHCFI%20Best%20Practices%20Report%20final2%202014%20Web.pdf

Gunawansa, Asanga, and Harn Wei Kua, "A Comparison of Climate Change Mitigation and Adaptation Strategies for the Construction Industries of Three Coastal Territories," *Sustainable Development*, Vol. 22, No. 1, January–February 2014.

Hallegatte, Stéphane, Ankur Shah, Robert Lempert, Casey Brown, and Stuart Gill, *Investment Decision Making Under Deep Uncertainty: Adaptation to Climate Change*, World Bank, September 2012.

Halofsky, Jessica E., David L. Peterson, Lara Y. Buluç, and Jason M. Ko, eds., *Climate Change Vulnerability and Adaptation for Infrastructure and Recreation in the Sierra Nevada*, U.S. Department of Agriculture, Forest Service, Pacific Southwest Research Station, September 2021.

Hammer, Stephen A., Lily Parshall, Robin Leichenko, Peter Vancura, and Marta Panero, "Energy," Ch. 8 in Cynthia Rosenzweig, William Solecki, Arthur DeGaetano, Megan O'Grady, Susan Hassol, and Paul Grabhorn, eds., *Responding to Climate Change in New York State: The ClimAID Integrated Assessment for Effective Climate Change Adaptation in New York State*, New York State Energy Research and Development Authority, No. 11-18, November 2011.

Han, Yu, Kevin Ash, Liang Mao, and Zhong-Ren Peng, "An Agent-Based Model for Community Flood Adaptation Under Uncertain Sea-Level Rise," *Climatic Change*, Vol. 162, 2020.

Han, Yu, and Pallab Mozumder, "Building-Level Adaptation Analysis Under Uncertain Sea-Level Rise," *Climate Risk Management*, Vol. 32, 2021.

Harrington, Elise, and Aileen Cole, "Typologies of Mutual Aid in Climate Resilience: Variation in Reciprocity, Solidarity, Self-Determination, and Resistance," *Environmental Justice*, Vol. 15, No. 3, 2022.

Harrison, Bob, "Reactivating Retirees for Police Service in Times of Crisis," *RAND Blog*, April 21, 2020. As of February 14, 2022:
https://www.rand.org/blog/2020/04/reactivating-retirees-for-police-service-in-times-of.html

HHS—*See* U.S. Department of Health and Human Services.

Hoverter, Sara P., *Adapting to Urban Heat: A Tool Kit for Governments*, Georgetown Climate Center, August 2012. As of September 7, 2022:
https://kresge.org/sites/default/files/climate-adaptation-urban-heat.pdf

Hsu, Angel, Glenn Sheriff, Tirthankar Chakraborty, and Diego Manya, "Disproportionate Exposure to Urban Heat Island Intensity Across Major US Cities," *Nature Communications*, Vol. 12, 2021.

HUD—*See* U.S. Department of Housing and Urban Development.

Hurd, Thomas, Gregory Beste, Rosemarie Grant, Paula Loomis, John Robinson, and Robert Thiele, *Disaster Assistance Handbook*, 3rd ed., American Institute of Architects, March 2017.

Institute for Tribal Environmental Professionals, "Climate Change Resources: Adaptation Planning Tool Kit," webpage, undated. As of August 15, 2022:
http://www7.nau.edu/itep/main/tcc/Resources/adaptation

Institute of Agriculture and Natural Resources, "Weather Ready Nebraska: Management Strategies," interactive graphic, University of Nebraska–Lincoln, Agritools, undated. As of August 15, 2022:
https://agritools.unl.edu/management-strategies/

Insurance Information Institute, *Fighting Wildfires with Innovation*, November 2019.

International Strategy for Disaster Reduction, *Hyogo Framework for Action 2005–2015: Building the Resilience of Nations and Communities to Disasters*, 2005.

Irfan, Umair, "The $5 Trillion Insurance Industry Faces a Reckoning. Blame Climate Change," Vox, October 15, 2021.

Jackson, Larry L., and Howard R. Rosenberg, "Preventing Heat-Related Illness Among Agricultural Workers," *Journal of Agromedicine*, Vol. 15, No. 3, July 2010.

Jacob, Klaus, George Deodatis, John Atlas, Morgan Whitcomb, Madeleine Lopeman, Olga Markogiannaki, Zackary Kennett, Aurelie Morla, Robin Leichenko, and Peter Vancura, "Transportation," Ch. 9 in Cynthia Rosenzweig, William Solecki, Arthur DeGaetano, Megan O'Grady, Susan Hassol, and Paul Grabhorn, eds., *Responding to Climate Change in New York State: The ClimAID Integrated Assessment for Effective Climate Change Adaptation in New York State*, New York State Energy Research and Development Authority, No. 11-18, November 2011.

Janowiak, Maria K., Daniel N. Dostie, Michael A. Wilson, Michael J. Kucera, R. Howard Skinner, Jerry L. Hatfield, David Hollinger, and Christopher W. Swanston, *Adaptation Resources for Agriculture: Responding to Climate Variability and Change in the Midwest and Northeast*, U.S. Department of Agriculture, October 2016.

Johnson, Denise, "Disaster Fraud Rampant After Floods, Hurricanes," *Claims Journal*, May 19, 2016.

Johnson, Kris A., Oliver E. J. Wing, Paul D. Bates, Joseph Fargione, Timm Kroeger, William D. Larson, Christopher C. Sampson, and Andrew M. Smith, "A Benefit-Cost Analysis of Floodplain Land Acquisition for US Flood Damage Reduction," *Nature Sustainability*, Vol. 3, No. 1, January 2020.

Joint Commission, "EC.02.03.05: The Organization Maintains Fire Safety Equipment and Fire Safety Building Features," undated. As of August 11, 2022:
https://www.jointcommission.org/resources/patient-safety-topics/the-physical-environment/fire-protection/

Jones, Alexi, "Cruel and Unusual Punishment: When States Don't Provide Air Conditioning in Prison," Prison Policy Initiative, June 18, 2019.

Jongman, Brenden, "Effective Adaptation to Rising Flood Risk," *Nature Communications*, Vol. 9, No. 1986, 2018.

Kane, Daniel A., Mark A. Bradford, Emma Fuller, Emily E. Oldfield, and Stephen A. Wood, "Soil Organic Matter Protects US Maize Yields and Lowers Crop Insurance Payouts Under Drought," *Environmental Research Letters*, Vol. 16, No. 4, March 2021.

Kim, Karl, and Lily Bui, "Learning from Hurricane Maria: Island Ports and Supply Chain Resilience," *International Journal of Disaster Risk Reduction*, Vol. 39, October 2019.

King, Sammy L., Murray K. Laubhan, Paul Tashjian, John Vradenburg, and Leigh Fredrickson, "Wetland Conservation: Challenges Related to Water Law and Farm Policy," *Wetlands*, Vol. 41, No. 54, 2021.

Kjellstrom, Tord, Nicolas Maître, Catherine Saget, Matthias Otto, and Tahmina Karimova, *Working on a Warmer Planet: The Impact of Heat Stress on Labour Productivity and Decent Work*, International Labour Organization, 2019.

Kramer, Melissa G., *Enhancing Sustainable Communities with Green Infrastructure: A Guide to Help Communities Better Manage Stormwater While Achieving Other Environmental, Public Health, Social, and Economic Benefits*, U.S. Environmental Protection Agency, Office of Sustainable Communities, October 2014. As of September 7, 2022:
https://www.epa.gov/smartgrowth/enhancing-sustainable-communities-green-infrastructure

Krausmann, Elizabeth, and Fesil Mushtaq, "A Qualitative Natech Damage Scale for the Impact of Floods on Selected Industrial Facilities," *Natural Hazards*, Vol. 46, 2008.

Kushner, Daniel, Jackie Ratner, and Jeff Schlegelmilch, "How to Build a More Resilient Energy Grid for the Future: Part 1," Columbia University Climate School, March 10, 2021.

Lafferty, David C., Ryan L. Sriver, Iman Haqiqi, Thomas W. Hertel, Klaus Keller, and Robert E. Nicholas, "Statistically Bias-Corrected and Downscaled Climate Models Underestimate the Adverse Effects of Extreme Heat on U.S. Maize Yields," *Communications Earth & Environment*, Vol. 2, No. 196, 2021.

Larsen, Larissa, "Urban Climate and Adaptation Strategies," *Frontiers in Ecology and the Environment*, November 1, 2015.

Lasater, Karen B., Linda H. Aiken, Douglas M. Sloane, Rachel French, Brendan Martin, Kyrani Reneau, Maryann Alexander, and Matthew D. McHugh, "Chronic Hospital Nurse Understaffing Meets COVID-19: An Observational Study," *BMJ Quality & Safety*, Vol. 30, No. 8, August 2021.

Lauland, Andrew, Benjamin Lee Preston, Kristin J. Leuschner, Michelle E. Miro, Liam Regan, Scott R. Stephenson, Rachel Steratore, Aaron Strong, Jonathan W. Welburn, and Jeffrey B. Wenger, *A Risk Assessment of National Critical Functions During COVID-19: Challenges and Opportunities*, Homeland Security Operational Analysis Center operated by the RAND Corporation, RR-A210-1, 2022. As of August 11, 2022:
https://www.rand.org/pubs/research_reports/RRA210-1.html

Lempert, Robert J., "Embedding (Some) Benefit-Cost Concepts into Decision Support Processes with Deep Uncertainty," *Journal of Benefit-Cost Analysis*, Vol. 5, No. 3, 2014.

Lempert, Robert, Jeffrey Arnold, Roger Pulwarty, Robert Lempert, Kate Gordon, Katherine Greig, Cat Hawkins Hoffman, Dale Sands, and Caitlin Werrell, "Reducing Risks Through Adaptation Actions," Ch. 28 in David Reidmiller, Christopher W. Avery, D. R. Easterling, K. E. Kunkel, K. L. M. Lewis, T. K. Maycock, and B. C. Stewart, eds., *Impacts, Risks, and Adaptation in the United States: Fourth National Climate Assessment*, Vol. II, U.S. Global Change Research Program, 2018. As of September 7, 2022:
https://nca2018.globalchange.gov/chapter/28/

Lempert, Robert J., and David G. Groves, "Identifying and Evaluating Robust Adaptive Policy Responses to Climate Change for Water Management Agencies in the American West," *Technological Forecasting and Social Change*, Vol. 77, No. 6, July 2010.

Lempert, Robert, James Syme, George Mazur, Debra Knopman, Garett Ballard-Rosa, Kacey Lizon, and Ifeanyi Edochie, "Meeting Climate, Mobility, and Equity Goals in Transportation Planning Under Wide-Ranging Scenarios: A Demonstration of Robust Decision Making," *Journal of the American Planning Association*, Vol. 86, No. 3, 2020.

Leonard, Matt, "Florida Manufacturing at Risk This Hurricane Season," Supply Chain Dive, June 3, 2019. As of October 18, 2022:
https://www.supplychaindive.com/news/florida-manufacturing-risk-hurricane-season/556020/

Levine, Carol, "The Concept of Vulnerability in Disaster Research," *Journal of Traumatic Stress*, Vol. 17, No. 5, 2004.

Linnenluecke, Martina K., Alexander Stathakis, and Andrew Griffiths, "Firm Relocation as Adaptive Response to Climate Change and Weather Extremes," *Global Environmental Change*, Vol. 21, No. 1, 2011.

Luber, George, and Michael McGeehin, "Climate Change and Extreme Heat Events," *American Journal of Preventive Medicine*, Vol. 35, No. 15, November 2008.

Macknick, J., S. Sattler, K. Averyt, S. Clemmer, and J. Rogers, "The Water Implications of Generating Electricity: Water Use Across the United States Based on Different Electricity Pathways Through 2050," *Environmental Research Letters*, Vol. 7, No. 14, 2012.

Maestro, Teresa, Alberto Garrido, and Maria Bielza, "Drought Insurance," Ch. 2.8, in Ana Iglesias, Dionysis Assimacopoulos, and Henny A. J. Van Lanen, eds., *Drought: Science and Policy*, Wiley, 2018.

Marchau, Vincent A. W. J., Warren E. Walker, Pieter J. T. M. Bloemen, and Steven W. Popper, *Decision Making Under Deep Uncertainty: From Theory to Practice*, Springer Nature, 2019.

Marmet, Barb, *Using Zoning Tools to Adapt to Sea Level Rise*, Virginia Coastal Policy Center, 2013.

Marteaux, Olivier, *Tomorrow's Railway and Climate Change Adaptation: Executive Report*, Rail Safety and Standards Board Limited, 2016.

Martin, Carlos, Gisela Campillo, Hilen Meirovich, and Jesus Navarrete, *Climate Change Mitigation & Adaptation Through Publically-Assisted Housing: Theoretical Framework for the IDB's Regional Policy Dialogue on Climate Change*, Inter-American Development Bank, September 2013.

McGranahan, Devan A., Paul W. Brown, Lisa A. Schulte, and John C. Tyndall, "A Historical Primer on the US Farm Bill: Supply Management and Conservation Policy," *Journal of Soil and Water Conservation*, Vol. 68, No. 3, May–June 2013.

Meerow, Sara, and Ladd Keith, "Planning for Extreme Heat: A National Survey of U.S. Planners," *Journal of the American Planning Association*, December 2021.

Mello, Steve, "More COPS, Less Crime," *Journal of Public Economics*, Vol. 172, April 2019.

Mercer, "US Healthcare Labor Market," white paper, 2021.

Milly, P. C. D., Julio Betancourt, Malin Falkenmark, Robert M. Hirsch, Zbigniew W. Kundzewicz, Dennis P. Lettenmaier, and Ronald J. Stouffer, "Stationarity Is Dead: Whither Water Management?" *Science*, Vol. 319, No. 5863, February 2008.

Miro, Michelle E., Andrew Lauland, Rahim Ali, Edward W. Chan, Richard H. Donohue, Liisa Ecola, Timothy R. Gulden, Liam Regan, Karen M. Sudkamp, Tobias Sytsma, Michael T. Wilson, and Chandler Sachs, *Assessing Risk to National Critical Functions as a Result of Climate Change*, RAND Corporation, RR-A1645-7, 2022. As of August 11, 2022:
https://www.rand.org/pubs/research_reports/RRA1645-7.html

Moritz, Max A., Enric Batllori, Ross A. Bradstock, A. Malcolm Gill, John Handmer, Paul Francis Hessburg, Justin Leonard, Sarah M. Mccaffrey, Dennis C. Odion, Tania Schoennage, and Alexandra D. Syphard, "Learning to Coexist with Wildfire," *Nature*, Vol. 515, No. 7525, 2014.

Mostyn, Sarah, et al., *The Workforce Crisis, and What Police Agencies Are Doing About It*, Police Executive Research Forum, September 2019.

Mount, Jeffrey, Ellen Hanak, Jay Lund, Paul Ullrich, Ken Baerenklau, Van Butsic, Caitrin Chappelle, Alvar Escriva-Bou, Graham Fogg, Greg Gartrell, Ted Grantham, Brian Gray, Sarge Green, Thomas Harter, Jelena Jezdimirovic, Yufang Jin, Henry McCann, Josué Medellín-Azuara, David Mitchell, Peter Moyle, Kurt Schwabe, Nathaniel Seavy, Scott Stephens, Leon Szeptycki, Barton "Buzz" Thompson, Joshua Viers, David Jassby, Alan Rhoades, Daniel Swain, and Zexuan Xu, *Managing Drought in a Changing Climate: Four Essential Reforms*, Public Policy Institute of California, September 2018.

Mulkern, Anne C., "Cliff Top Trains Could Race into Tunnels to Avoid Rising Seas," ClimateWire, January 25, 2022.

Multi-Hazard Mitigation Council, *Natural Hazard Mitigation Saves: 2019 Report*, National Institute of Building Sciences, 2019.

Munir, Nur Al Anshari, Yogasmana Al Mustafa, and Fransileo Siagian, "Analysis of the Effect of Ambient Temperature and Loading on Power Transformers Ageing (Study Case of 3rd Power Transformer in Cikupa Substation)," *2019 2nd International Conference on High Voltage Engineering and Power Systems (ICHVEPS)*, Institute of Electrical and Electronics Engineers, 2019, pp. 1–6.

Nabeel, Ismail, Yohama Caraballo-Arias, William Brett Perkison, Ronda B. McCarthy, Pouné Saberi, Manijeh Berenji, Rose H. Goldman, Jasminka Goldoni Laestadius, Rosemary K. Sokas, Rupali Das, et al., "Proposed Mitigation and Adaptation Strategies Related to Climate Change: Guidance for OEM Professionals," *Journal of Occupational and Environmental Medicine*, Vol. 63, No. 9, September 2021.

Nateghi, Roshanak, "Multi-Dimensional Infrastructure Resilience Modeling: An Application to Hurricane-Prone Electric Power Distribution Systems," *IEEE Access*, Vol. 6, January 2018.

National Academies of Sciences, Engineering, and Medicine, *The Future of Nursing 2020–2030: Charting a Path to Achieve Health Equity*, National Academies Press, 2021.

National Cooperative Highway Research Program, *Incorporating the Costs and Benefits of Adaptation Measures in Preparation for Extreme Weather Events and Climate Change—Guidebook*, National Academies Press, 2020. As of February 7, 2022:
https://nap.nationalacademies.org/catalog/25744/
incorporating-the-costs-and-benefits-of-adaptation-measures-in-preparation

National Emergency Management Association, "Emergency Management Assistance Compact: The All Hazards National Mutual Aid System," undated. As of September 9, 2022:
https://www.emacweb.org/

National Fish and Wildlife Foundation, "Coastal Resilience Evaluation and Siting Tool (CREST)," webpage, undated.

National Fish, Wildlife, and Plants Climate Adaptation Network, *Advancing the National Fish, Wildlife, and Plants Climate Adaptation Strategy into a New Decade*, 2021. As of September 7, 2022:
https://www.fishwildlife.org/application/files/4216/1161/3356/Advancing_Strategy_Report_FINAL.pdf

National Hog Farmer, "10 Things to Prepare Your Farm for Disaster," September 2017.

National Integrated Drought Information System, "Drought Impacts on Manufacturing," Drought.gov, undated. As of September 8, 2022:
https://www.drought.gov/sectors/manufacturing

National Oceanographic and Atmospheric Administration, "Health Care Facilities Maintain Indoor Air Quality Through Smoke and Wildfires," webpage, August 15, 2017.

National Research Council, *Developing a Framework for Measuring Community Resilience: Summary of a Workshop*, National Academies Press, 2015.

National Weather Service Instruction 10-1605, *Storm Data Preparation*, July 26, 2021.

Nicholls, Robert J., and Richard S. J. Tol, "Impacts and Responses to Sea-Level Rise: A Global Analysis of the SRES Scenarios over the Twenty-First Century," *Philosophical Transactions of the Royal Society A: Mathematical, Physical and Engineering Sciences*, Vol. 364, No. 1841, April 15, 2006.

NOAA—*See* National Oceanographic and Atmospheric Administration.

Noble, Ian R., Yuri A. Anokhin, JoAnn Carmin, Dieudonne Goudou, Felino P. Lansigan, Balgis Osman-Elasha and Alicia Villamizar, "Adaptation Needs and Options," Ch. 14 in C. B. Field, V. R. Barros, D. J. Dokken, K. J. Mach, M. D. Mastrandrea, T. E. Bilir, M. Chatterjee, K. L. Ebi, Y. O. Estrada, R. C. Genova, B. Girma, E. S. Kissel, A. N. Levy, S. MacCracken, P. R. Mastrandrea, and L. L. White, eds., *Climate Change 2014: Impacts, Adaptation, and Vulnerability*, Part A: *Global and Sectoral Aspects*, Contribution of Working Group II to the Fifth Assessment Report of the Intergovernmental Panel on Climate Change, Cambridge University Press, 2014.

Northern Institute of Applied Climate Science, "Adaptation Workbook," homepage, undated. As of August 15, 2022:
https://adaptationworkbook.org/

Notre Dame Global Adaptation Initiative, "Urban Adaptation Assessment," webpage, undated. As of August 15, 2022:
https://gain-uaa.nd.edu/

Occupational Safety and Health Administration, "U.S. Department of Labor Announces Enhanced, Expanded Measures to Protect Workers from Hazards of Extreme Heat, Indoors and Out," press release, U.S. Department of Labor, September 20, 2021.

OECD—*See* Organisation for Economic Co-operation and Development.

O'Grady, Megan, Colleen Moore, Joel B. Smith, Heather Hosterman, Christine Teter, Alexis St. Juliana, Michael Duckworth, and Diane Callow, "U.S. Department of Housing and Urban Development Community Resilience Toolkit," February 2022. As of September 7, 2022:
https://files.hudexchange.info/resources/documents/HUD-Community-Resilient-Toolkit.pdf

Oregon State University, "AgBiz Logic," homepage, 2022. As of August 15, 2022:
https://www.agbizlogic.com/

Organisation for Economic Co-operation and Development, "Climate-Resilient Infrastructure," OECD Environment Policy Paper 14, 2018.

Organisation for Economic Co-operation and Development, "Building Agricultural Resilience to Extreme Floods in the United States," Ch. 11 in Organisation for Economic Co-operation and Development, Building Agricultural Resilience to Natural Hazard-Induced Disasters: Insights from Country Case Studies, 2021. As of August 12, 2022:
https://www.oecd-ilibrary.org/sites/bd6c9ca1-en/index.html?itemId=/content/component/bd6c9ca1-en

OSHA—See Occupational Safety and Health Administration.

Pachon, Carlos, Anne Dailey, and Ellen Treimel, "Climate Change Adaptation Technical Fact Sheet: Landfills and Containment as an Element of Site Remediation," U.S. Environmental Protection Agency, Office of Superfund Remediation and Technology, May 2014.

Pal, Govinda, Thaneswer Patel, and Trishita Banik, "Effect of Climate Change Associated Hazards on Agricultural Workers and Approaches for Assessing Heat Stress and Its Mitigation Strategies—Review of Some Research Significances," International Journal of Current Microbiology and Applied Sciences, Vol. 10, No. 2, 2021.

Palin, Erika J., Irina Stipanovic Oslakovic, Kenneth Gavin, and Andrew Quinn, "Implications of Climate Change for Railway Infrastructure," WIREs Climate Change, Vol. 12, No. 5, 2021.

Pan, Xiaodan, Martin Dresner, Benny Mantin, and Jun A. Zhang, "Pre-Hurricane Consumer Stockpiling and Post-Hurricane Product Availability: Empirical Evidence from Natural Experiments," Production and Operations Management, Vol. 29, No. 10, October 2020.

Panteli, Mathaios, and Pierluigi Mancarella, "Influence of Extreme Weather and Climate Change on the Resilience of Power Systems: Impacts and Possible Mitigation Strategies," Electric Power Systems Research, Vol. 127, October 2015.

Parker, Lauren E., Andrew J. McElrone, Steven M. Ostoja, and Elisabeth J. Forrestel, "Extreme Heat Effects on Perennial Crops and Strategies for Sustaining Future Production," Plant Science, Vol. 295, June 2020.

Paul, Jomon Aliyas, and Govind Hariharan, "Location-Allocation Planning of Stockpiles for Effective Disaster Mitigation," Annals of Operations Research, Vol. 196, No. 1, July 2012.

Paul, Jomon Aliyas, and Leo MacDonald, "Optimal Location, Capacity and Timing of Stockpiles for Improved Hurricane Preparedness," International Journal of Production Economics, Vol. 174, 2016.

Peacock, Walter Gillis, and Rahmawati Husein, The Adoption and Implementation of Hazard Mitigation Policies and Strategies by Coastal Jurisdictions in Texas: The Planning Survey Results, Texas A&M University, Hazard Reduction and Recovery Center, December 2011.

Pettit, Timothy J., Keely L. Croxton, and Joseph Fiksel, "Ensuring Supply Chain Resilience: Development and Implementation of an Assessment Tool," Journal of Business Logistics, Vol. 34, No. 1, March 2013.

PolicyLink and the USC Equity Research Institute, "National Equity Atlas," homepage, undated. As of August 12, 2022:
https://nationalequityatlas.org/

Pörtner, Hans-O., Debra C. Roberts, Elvira S. Poloczanska, Katja Mintenbeck, Melinda Tignor, Andrés Alegría, M. Craig, Stephanie Langsdorf, Sina Löschke, Vincent Möller, and Andrew Okem, eds., "Summary for Policymakers," in Hans-O. Pörtner, Debra C. Roberts, Elvira S. Poloczanska, Katja Mintenbeck, Melinda Tignor, Andrés Alegría, M. Craig, Stephanie Langsdorf, Sina Löschke, Vincent Möller, and Andrew Okem, eds., Climate Change 2022: Impacts, Adaptation, and Vulnerability, Contribution of Working Group II to the Sixth Assessment Report of the Intergovernmental Panel on Climate Change, Cambridge University Press, 2022. As of September 7, 2022:
https://www.ipcc.ch/report/sixth-assessment-report-working-group-ii/

Qin, Rongshui, Nima Khakzad, and Jiping Zhu, "An Overview of the Impact of Hurricane Harvey on Chemical and Process Facilities in Texas," *International Journal of Disaster Risk Reduction*, Vol. 45, May 2020.

Quinton, Sophie, "Lack of Federal Firefighters Hurts California Wildfire Response," webpage, Pew Charitable Trusts, July 14, 2021.

Rattanachot, Wit, Yuhong Wang, Dan Chong, and Suchatvee Suwansawas, "Adaptation Strategies of Transport Infrastructures to Global Climate Change," *Transport Policy*, Vol. 41, July 2015.

Ready.gov, "Citizen Corps," website, undated. As of February 16, 2022:
https://www.ready.gov/citizen-corps

Reams, Margaret A., Terry K. Haines, Cheryl R. Renner, Michael W. Wascoma, and Harish Kingre, "Goals, Obstacles and Effective Strategies of Wildfire Mitigation Programs in the Wildland-Urban Interface," *Forest Policy and Economics*, Vol. 7, 2005.

Reef Resilience Network, "Community-Based Climate Adaptation," webpage, undated. As of August 15, 2022:
https://reefresilience.org/management-strategies/community-based-climate-adaptation/

Reniers, Genserik, Nima Khakzad, Valerio Cozzani, and Faisal Khane, "The Impact of Nature on Chemical Industrial Facilities: Dealing with Challenges for Creating Resilient Chemical Industrial Parks," *Journal of Loss Prevention in the Process Industries*, Vol. 56, November 2018.

Rice, James B., Jr., Kai Trepte, and Ken Cottrill, "Port Mapper: Preparing for the Future," Eno Center for Transportation, June 26, 2013. As of October 18, 2022:
https://www.enotrans.org/article/port-mapper-preparing-future/

Robert, Kates W., Thomas M. Parris, and Anthony A. Leiserowitz, "What Is Sustainable Development? Goals, Indicators, Values, and Practice," *Environment: Science and Policy for Sustainable Development*, Vol. 47, No. 3, 2005.

Romero-Lankao, Patricia, Joel B. Smith, Debra J. Davidson, Noah S. Diffenbaugh, Patrick L. Kinney, Paul Kirshen, Paul Kovacs, and Lourdes Villers Ruiz, "North America," in V. R. Barros, C. B. Field, D. J. Dokken, M. D. Mastrandrea, K. J. Mach, T. E. Bilir, M. Chatterjee, K. L. Ebi, Y. O. Estrada, R. C. Genova, B. Girma, E. S. Kissel, A. N. Levy, S. MacCracken, P. R. Mastrandrea, and L. L. White, eds., *Climate Change 2014: Impacts, Adaptation, and Vulnerability*, Part B: *Regional Aspects*, Contribution of Working Group II to the Fifth Assessment Report of the Intergovernmental Panel on Climate Change, Cambridge University Press, 2014.

Roth, Jeffrey A., and Joseph F. Ryan, "The COPS Program After 4 Years—National Evaluation," U.S. Department of Justice, National Institute of Justice, August 2000.

Rowlinson, Steve, and Yunyan Andrea Jia, "Application of the Predicted Heat Strain Model in Development of Localized, Threshold-Based Heat Stress Management Guidelines for the Construction Industry," *Annals of Occupational Hygiene*, Vol. 58, No. 3, 2014.

Scata, Joel, "FEMA Moves to Reform Flood Insurance Program," Natural Resources Defense Council, October 14, 2021.

Schulz, Kevin, "10 Things to Prepare Your Farm for Disaster," webpage, National Hog Farmer, September 7, 2017.

Shi, Linda, "From Progressive Cities to Resilient Cities: Lessons from History for New Debates in Equitable Adaptation to Climate Change," *Urban Affairs Review*, Vol. 57, No. 5, 2021.

Shirzaei, Manoochehr, Mostafa Khoshmanesh, Chandrakanta Ojhac, Susanna Werth, Hannah Kerner, Grace Carlson, Sonam Futi Sherpa, Guang Zhai, and Jui-ChiLeea, "Persistent Impact of Spring Floods on Crop Loss in U.S. Midwest," *Weather and Climate Extremes*, Vol. 34, December 2021.

Social and Environmental Research Institute, Inc., "Vulnerability, Consequences, and Adaptation Planning Scenarios (VCAPS)," webpage, undated. As of August 15, 2022:
http://www.vcapsforplanning.org/

Southern California Association of Governments, "Adaptation and Resilience Planning: For Providers of Public Transportation," webpage, undated. As of August 15, 2022:
https://scag.ca.gov/transit-adaptation-and-resilience-planning

Stagrum, Anna Eknes, Erlend Andenaes, Tore Kvande, and Jardar Lohne, "Climate Change Adaptation Measures for Buildings—A Scoping Review," *Sustainability*, Vol. 12, No. 5, February 25, 2020.

Stamos, Iraklis, Evangelos Mitsakis, and Josep Maria Salanova Grau, "Roadmaps for Adaptation Measures of Transportation to Climate Change," *Transportation Research Record*, No. 2532, 2015.

Stephens, Jennie C., Elizabeth J. Wilson, Tarla R. Peterson, and James Meadowcroft, "Getting Smart? Climate Change and the Electric Grid," *Challenges*, Vol. 12, No. 2, 2013.

Svoboda, Mark D., Brian A. Fuchs, Chris C. Poulsen, and Jeff R. Nothwehr, "The Drought Risk Atlas: Enhancing Decision Support for Drought Risk Management in the United States," *Journal of Hydrology*, Vol. 526, July 2015.

Taylor, Ciji, "Gear Up for the 2021 Hurricane Season: Prepare and Recover with USDA," U.S. Department of Agriculture, Farmers.gov, July 2, 2021.

Taylor, Sofia, and Line A. Roald, "A Framework for Risk Assessment and Optimal Line Upgrade Selection to Mitigate Wildfire Risk," paper submitted for the 22nd Power Systems Computation Conference, 2021.

Tian, Zhongbei, Ning Zhao, Stuart Hillmansen, Clive Roberts, Trevor Dowens, and Colin Kerr, "SmartDrive: Traction Energy Optimization and Applications in Rail Systems," *IEEE Transactions on Intelligent Transportation Systems*, Vol. 20, No. 7, 2019.

Tigchelaar, Michelle, David S. Battisti, and June T. Spector, "Work Adaptations Insufficient to Address Growing Heat Risk for U.S. Agricultural Workers," *Environmental Research Letters*, Vol. 15, No. 9, 2020.

Titus, James G., *Rolling Easements*, U.S. Environmental Protection Agency, June 2011. As of October 21, 2022: https://www.epa.gov/sites/default/files/documents/rollingeasementsprimer.pdf

Toplis, Caroline, Murray Kidnie, April Marchese, Cristina Maruntu, Helen Murphy, Robin Sebille, and Stephen Thomson, *International Climate Change Adaptation Framework for Road Infrastructure*, PIARC–World, 2015. As of February 7, 2022: https://www.piarc.org/en/order-library/23517-en-International%20climate%20change%20adaptation%20 framework%20for%20road%20infrastructure.htm

Troy, Austin, J. Morgan Grove, and Jarlath O'Neill-Dunne, "The Relationship Between Tree Canopy and Crime Rates Across an Urban-Rural Gradient in the Greater Baltimore Region," *Landscape and Urban Planning*, Vol. 106, No. 3, 2012.

Truelove, Heather Barnes, Amanda R. Carrico, Elke U. Weber, Kaitlin Toner Raimi, and Michael P. Vandenbergh, "Positive and Negative Spillover of Pro-Environmental Behavior: An Integrative Review and Theoretical Framework," *Global Environmental Change*, Vol. 29, November 2014.

Tully, Kate, Keryn Gedan, Rebecca Epanchin-Niell, Aaron Strong, Emily S. Bernhardt, Todd BenDor, Molly Mitchell, John Kominoski, Thomas E. Jordan, Scott C. Neubauer, and Nathaniel B. Weston, "The Invisible Flood: The Chemistry, Ecology, and Social Implications of Coastal Saltwater Intrusion," *BioScience*, Vol. 69, No. 5, May 2019.

Turek-Hankins, Lynée L., Erin Coughlan de Perez, Giulia Scarpa, Raquel Ruiz-Diaz, Patricia Nayna Schwerdtle, Elphin Tom Joe, Eranga K. Galappaththi, Emma M. French, Stephanie E. Austin, Chandni Singh, et al., "Climate Change Adaptation to Extreme Heat: A Global Systematic Review of Implemented Action," *Oxford Open Climate Change*, Vol. 1, No. 1, 2021.

Twaddell, Hannah, Alanna McKeeman, Michael Grant, Jessica Klion, Uri Avin, Kate Ange, and Mike Callahan, *Supporting Performance-Based Planning and Programming Through Scenario Planning*, U.S. Department of Transportation, Federal Highway Administration, June 2016. As of September 7, 2022: https://www.fhwa.dot.gov/planning/scenario_and_visualization/scenario_planning/ scenario_planning_guidebook/

UNEP—*See* United Nations Environment Programme.

United Nations Environment Programme, "A Practical Guide to Climate-Resilient Buildings & Communities," 2021.

United Nations Environment Programme, Secretariat of the Minamata Convention on Mercury, and Rotterdam Secretariat of the Basel, "Chemicals, Wastes and Climate Change: Interlinkages and Potential for Coordinated Action," 2021.

University of Nebraska–Lincoln, "National Drought Mitigation Center," homepage, undated. As of August 15, 2022: https://drought.unl.edu/

U.S. Army Corps of Engineers, "Natural Infrastructure Opportunities Tool—Connecting Resources to Needs," webpage, ArcGIS Online, undated. September 7, 2022:
https://www.arcgis.com/apps/MapSeries/index.html?appid=18079f5b628b4a7bb52acbe089d80886

U.S. Climate Resilience Toolkit, homepage, undated. As of August 30, 2022:
http://toolkit.climate.gov

USDA—*See* U.S. Department of Agriculture.

U.S. Department of Agriculture, "Compendium of Adaptation Approaches," webpage, U.S. Forest Service Climate Change Resource Center, undated a. As of August 15, 2022:
https://www.fs.usda.gov/ccrc/climate-projects/adaptation-approaches

U.S. Department of Agriculture, "Farming the Floodplain: Trade-Offs and Opportunities," webpage, undated b. As of August 12, 2022:
https://www.climatehubs.usda.gov/hubs/northeast/topic/farming-floodplain-trade-offs-and-opportunities

U.S. Department of Agriculture, *Action Plan for Climate Adaptation and Resilience*, August 2021a.

U.S. Department of Agriculture, "USDA, DOI, and FEMA Jointly Establish New Wildland Fire Mitigation and Management Commission," press release, December 17, 2021b.

U.S. Department of Energy, "Disaster Resistance," webpage, Office of Energy Efficiency & Renewable Energy, Building America Solution Center, undated. As of August 15, 2022:
https://basc.pnnl.gov/disaster-resistance

U.S. Department of Energy, *Climate Change and the U.S. Energy Sector: Regional Vulnerabilities and Resilience Solutions*, October 2015. As of December 14, 2021:
https://toolkit.climate.gov/reports/
climate-change-and-us-energy-sector-regional-vulnerabilities-and-resilience-solutions

U.S. Department of Energy, "Reimagining and Rebuilding America's Energy Grid," webpage, Energy.gov, June 10, 2021. As of October 18, 2022:
https://www.energy.gov/articles/reimagining-and-rebuilding-americas-energy-grid

U.S. Department of Health and Human Services, *Primary Protection: Enhancing Health Care Resilience for a Changing Climate*, December 2014.

U.S. Department of Health and Human Services, Assistant Secretary for Preparedness and Response, Technical Resources, Assistance Center, and Information Exchange, *The Exchange*, No. 10, 2020. As of September 9, 2022:
https://www.hsdl.org/?abstract&did=836243

U.S. Department of Housing and Urban Development, "Housing and Urban Development Climate Change Adaptation Plan," October 2014.

U.S. Department of Housing and Urban Development, Office of Economic Resilience, "Green Infrastructure and the Sustainable Communities Initiative," March 2015.

U.S. Department of Housing and Urban Development, "Department of Housing and Urban Development 2021 Climate Adaptation Plan," September 2021.

U.S. Department of Housing and Urban Development, "Community Resilience Toolkit," "Supporting Local Climate Action," webpage, HUD Exchange website, January 31, 2022. As of October 18, 2022:
https://www.hudexchange.info/programs/supporting-local-climate-action

U.S. Department of the Navy, "Climate Change: Installation Adaptation and Resilience Handbook," Naval Facilities Command, January 2017.

U.S. Environmental Protection Agency, "Agriculture and Natural Events and Disasters," webpage, undated a. As of August 15, 2022:
https://www.epa.gov/agriculture/agriculture-and-natural-events-and-disasters

U.S. Environmental Protection Agency, "Build Wildfire Resilience," webpage, undated b. As of August 15, 2022:
https://www.epa.gov/waterutilityresponse/build-wildfire-resilience

U.S. Environmental Protection Agency, "Climate Change Adaptation Resource Center (ARC-X): Climate Impacts on Water Utilities," webpage, undated c. As of August 15, 2022:
https://www.epa.gov/arc-x/climate-impacts-water-utilities

U.S. Environmental Protection Agency, *Climate Change Handbook for Regional Water Planning*, undated d. As of August 15, 2022:
https://www.epa.gov/arc-x/climate-change-handbook-regional-water-planning-pdf

U.S. Environmental Protection Agency, "Creating Resilient Water Utilities (CRWU)," webpage, undated e. As of August 15, 2022:
https://www.epa.gov/crwu

U.S. Environmental Protection Agency, "EJScreen: Environmental Justice Screening and Mapping Tool," webpage, undated f. As of August 15, 2022:
https://www.epa.gov/ejscreen

U.S. Environmental Protection Agency, "Storm Water Management Model (SWMM)," webpage, undated g. As of August 15, 2022:
https://www.epa.gov/water-research/storm-water-management-model-swmm

U.S. Environmental Protection Agency, *Synthesis of Adaptation Options for Coastal Areas*, Climate Ready Estuaries, 2009. As of September 7, 2022:
https://www.epa.gov/sites/default/files/2014-04/documents/cre_synthesis_1-09.pdf

U.S. Environmental Protection Agency, "Office of Solid Waste and Emergency Response, Climate Change Adaptation Implementation Plan," June 2014a.

U.S. Environmental Protection Agency, "Flood Resilience Checklist," July 2014b. As of August 15, 2022:
https://www.epa.gov/sites/default/files/2014-07/documents/flood-resilience-checklist.pdf

U.S. Environmental Protection Agency, "Flood Resilience: A Basic Guide for Water and Wastewater Utilities," September 2014c. As of August 15, 2022:
https://www.epa.gov/sites/default/files/2015-08/documents/flood_resilience_guide.pdf

U.S. Environmental Protection Agency, "Strategies for Climate Change Adaptation," Climate Change Adaptation Resource Center (ARC-X), March 23, 2016.

U.S. Environmental Protection Agency, "Adaptation Strategies Guide for Water Utilities" Climate Ready Water Utilities, 2017a. As of September 7, 2022:
https://www.epa.gov/sites/default/files/2015-04/documents/
updated_adaptation_strategies_guide_for_water_utilities.pdf

U.S. Environmental Protection Agency, *Smart Growth Fixes for Climate Adaptation and Resilience: Changing Land Use and Building Codes and Policies to Prepare for Climate Change*, January 2017b.

U.S. Environmental Protection Agency, *Drought Response and Recovery: A Basic Guide for Water Utilities*, August 2018. As of September 7, 2022:
https://www.epa.gov/waterutilityresponse/drought-response-and-recovery-guide-water-utilities

U.S. Environmental Protection Agency, "Climate Resilience Evaluation and Awareness Tool (CREAT) Risk Assessment Application for Water Utilities," webpage, 2022a. As of August 15, 2022:
https://www.epa.gov/crwu/
climate-resilience-evaluation-and-awareness-tool-creat-risk-assessment-application-water

U.S. Environmental Protection Agency, "National Stormwater Calculator," webpage, 2022b. As of August 15, 2022:
https://www.epa.gov/water-research/national-stormwater-calculator

U.S. Environmental Protection Agency, "Resilient Strategies Guide for Water Utilities," webpage, 2022c. As of August 15, 2022:
https://www.epa.gov/crwu/resilient-strategies-guide-water-utilities#/
?region=101&utilityType=4&utilitySize=1315&assets=&priorities=&strategies=&fundingSources=

U.S. Fire Administration, *Fire Department Preparedness for Extreme Weather Emergencies and Natural Disasters*, USFA-TR-162, April 2008.

U.S. Forest Service, "Innovating Wildfire Insurance," 2021.

USGCRP—*See* U.S. Global Change Research Program.

U.S. Global Change Research Program, "Reports & Resources," webpage, undated. As of September 7, 2022:
https://www.globalchange.gov/browse

U.S. Government Accountability Office, *Actions Needed to Address Deployment and Staff Development Challenges*, GAO-20-360, May 2020.

U.S. Government Accountability Office, "National Flood Insurance Program: Congress Should Consider Updating the Mandatory Purchase Requirement," webpage, July 2021. As of September 9, 2022: https://www.gao.gov/products/gao-21-578

Verschuur, J., E. E. Koks, and J. W. Hall, "Port Disruptions Due to Natural Disasters: Insights into Port and Logistics Resilience," *Transportation Research Part D: Transport and Environment*, Vol. 85, 2020.

Vine, Edward, "Adaptation of California's Electricity Sector to Climate Change," *Climatic Change*, No. 111, October 6, 2011.

Ward, Philip J., Brenden Jongman, Jeroen C. J. H. Aerts, Paul D. Bates, Wouter J. W. Botzen, Andres Diaz Loaiza, Stephane Hallegatte, Jarl M. Kind, Jaap Kwadijk, Paolo Scussolini, and Hessel C. Winsemius, "A Global Framework for Future Costs and Benefits of River-Flood Protection in Urban Areas," *Nature Climate Change*, Vol. 7, 2017.

White House, "Biden Administration Mobilizes to Protect Workers and Communities from Extreme Heat," fact sheet, September 20, 2021.

Wickham, Elliot D., Deborah Bathke, Tarik Abdel-Monem, Tonya Bernadt, Denise Bulling, Lisa Pytlik-Zillig, Crystal Stiles, and Nicole Wall, "Conducting a Drought-Specific THIRA (Threat and Hazard Identification and Risk Assessment): A Powerful Tool for Integrating All-Hazard Mitigation and Drought Planning Efforts to Increase Drought Mitigation Quality," *International Journal of Disaster Risk Reduction*, Vol. 39, October 2019.

Wiener, Sarah S., Nora L. Álvarez-Berríos, and Angela B. Lindsey, "Opportunities and Challenges for Hurricane Resilience on Agricultural and Forest Land in the U.S. Southeast and Caribbean," *Sustainability*, Vol. 12, No. 4, 2020.

Wildland Fire Leadership Council, "Wildland Fire Leadership Council," Forests and Rangelands website, undated. https://www.forestsandrangelands.gov/leadership/

World Commission on Environment and Development, *Our Common Future*, New York: Oxford University Press, 1987.

Wuebbles, Donald J., David W. Fahey, Kathy A. Hibbard, David J. Dokken, Brooke C. Stewart, and Thomas K. Maycock, eds., *Fourth National Climate Assessment*, Vol. I: *Climate Science Special Report*, U.S. Global Change Research Program, 2017.

Yesudian, Aaron N., and Richard J. Dawson, "Global Analysis of Sea Level Rise Risk to Airports," *Climate Risk Management*, Vol. 31, 2021.